Embedded Systems

Series editors
Nikil D. Dutt, Irvine, CA, USA
Grant Martin, Santa Clara, CA, USA
Peter Marwedel, Dortmund, Germany

This Series addresses current and future challenges pertaining to embedded hardware, software, specifications and techniques. Titles in the Series cover a focused set of embedded topics relating to traditional computing devices as well as high-tech appliances used in newer, personal devices, and related topics. The material will vary by topic but in general most volumes will include fundamental material (when appropriate), methods, designs and techniques.

More information about this series at http://www.springer.com/series/8563

Anup Kumar Das • Akash Kumar
Bharadwaj Veeravalli • Francky Catthoor

Reliable and Energy Efficient Streaming Multiprocessor Systems

 Springer

Anup Kumar Das
Electrical and Computer
Engineering Department
Drexel University, Bossone Research
Enterprise Center
Philadelphia, PA, USA

Bharadwaj Veeravalli
Department of ECE
National University of Singapore
Singapore, Singapore

Akash Kumar
Chair for Processor Design
Dresden University of Technology
Dresden, Germany

Francky Catthoor
IMEC
Heverlee, Belgium

ISSN 2193-0155   ISSN 2193-0163   (electronic)
Embedded Systems
ISBN 978-3-319-88766-1   ISBN 978-3-319-69374-3   (eBook)
https://doi.org/10.1007/978-3-319-69374-3

Printed on acid-free paper

This Springer imprint is published by Springer Nature
The registered company is Springer International Publishing AG
The registered company address is: Gewerbestrasse 11, 6330 Cham, Switzerland

This book is dedicated to those who continue the pursuit of knowledge, despite the obstacles life presents. You hold the key to our future. May God bless you with determination on your journey. Never let those doubts or negativity ruin your spirit. Be steadfast in your quest for knowledge. God Bless.

Preface

As the performance demands of applications (e.g., multimedia) are growing, multiple processing cores are integrated together to form multiprocessor systems. Energy minimization is a primary optimization objective for these systems. An emerging concern for designs at deep-submicron technology nodes (65 nm and below) is the lifetime reliability, as escalating power density and hence temperature variation continues to accelerate wear-out leading to a growing prominence of device defects. As such, reliability and energy need to be incorporated in the multiprocessor design methodology, addressing two key aspects:

- lifetime amelioration, i.e., improving the lifetime reliability through energy- and performance-aware intelligent task mapping
- graceful degradation, i.e., determining the task mapping for different fault-scenarios while minimizing the energy consumption and providing a graceful performance degradation

In this book, a platform-based design methodology is first proposed to minimize temperature-related wear-outs. Fundamental to this methodology is a temperature model that predicts the temperature of a core incorporating not only its dependency on the voltage and frequency of operation (temporal effect), but also its dependency on the temperature of the surrounding cores (spatial effect). The proposed temperature model is integrated in a gradient-based fast heuristic that controls the voltage and frequency of the cores to limit the average and peak temperature leading to a longer lifetime, simultaneously minimizing the energy consumption.

A design flow is then proposed as a part of the hardware-software co-design methodology to determine the minimum number of cores and the size of the FPGA fabric of a reconfigurable multiprocessor system. The objective is to maximize the lifetime reliability of the cores while satisfying a given area, performance, and energy budget. The proposed flow incorporates individual as well as concurrent applications with different performance requirements and thermal behaviors. While the existing studies determine platform architecture for energy and area minimization, this is the first approach for reconfigurable multiprocessor system design considering lifetime reliability together with multi-application use-cases.

To provide graceful performance degradation in the presence of faults, a reactive fault-tolerance technique is also proposed that explores different task mapping alternatives to minimize energy consumption while guaranteeing throughput for all processor fault-scenarios. Directed acyclic graphs (DAGs) and synchronous data flow graphs (SDFGs) are used to model applications making the proposed methodology applicable to streaming multimedia and non-multimedia applications. Fundamental to this approach is a novel scheduling algorithm based on self-timed execution, which minimizes both the schedule storage overhead and run-time schedule construction overhead. Unlike the existing approaches which consider task mapping only, the proposed technique considers task mapping and scheduling in an integrated manner, achieving significant improvement with respect to these state-of-the-art approaches.

Finally, an adaptive run-time manager is designed for lifetime amelioration of multiprocessor systems by managing thermal variations, both within (intra) and across (inter) applications. Core to this approach is a reinforcement learning algorithm, which interfaces with the on-board thermal sensors and controls the voltage and frequency of operation and the thread-to-core affinity over time. This approach is built on top of the design-time analysis to minimize learning time, address run-time variability, and support new applications, making the overall book objective to provide a complete and systematic design solution for reliable and energy-efficient multiprocessor systems.

Texas, United States   Anup Kumar Das
Dresden, Germany   Akash Kumar
Singapore, Singapore   Bharadwaj Veeravalli
Heverlee, Belgium   Francky Catthoor

Contents

Acronyms

ALU	Arithmetic logic unit
ASIC	Application-specific integrated circuit
CMOS	Complementary metal-oxide-semiconductor
DAG	Directed acyclic graph
DMR	Double modular redundancy
DSE	Design space exploration
DVFS	Dynamic voltage and frequency scaling
EM	Electromigration
FET	Field effect transistor
FFT	Fast Fourier transform
FPGA	Field programmable gate array
HCI	Hot carrier injection
HSDFG	Homogeneous synchronous data flow graph
ILP	Integer linear programming
ISA	Instruction set architecture
ITRS	International technology roadmap for semiconductors
JPEG	Joint photographic experts group
MOSFET	Metal-oxide-semiconductor field-effect transistor
MP3	MPEG-1 or MPEG-2 audio layer III
MPSoC	Multiprocessor system-on-chip
MPEG	Moving picture experts group
MTBF	Mean-time-between-failures
MTTF	Mean-time-to-failure
NBTI	Negative-bias temperature instability
NMOS	n-channel MOSFET
NoC	Network-on-chip
NRE	Non-recurrent engineering
NUMA	Non-uniform memory access
PMOS	p-channel MOSFET
PVT	Process, voltage, temperature
QPI	Quick-path interconnect

RTL	Register transfer language
SDFG	Synchronous data flow graph
SoC	System-on-chip
TDDB	Time-dependent gate oxide breakdown
TMR	Triple modular redundancy
TTFF	Time-to-first-failure

Chapter 1
Introduction

1.1 Trends in Multiprocessor Systems Design

Multiprocessor systems have evolved over the past decades, triggered by innovations in transistor scaling and integration, multiprocessor design and system integration. This section summarizes these trends.

1.1.1 Trends in Transistor Scaling

Technology scaling principles introduced in 1974 [21] have revolutionized the semiconductor industry, providing a roadmap for systematic and predictable improvement in transistor density, switching speed, and power dissipation, with scaled transistor feature size. Ever since these principles were introduced, transistor feature size has reduced by $\sqrt{2}$ approximately every 18 months, starting from a CMOS gate length of 1 mm in 1974 [21] to commercial 22 nm designs available as early as in 2012 [32]. Current technological breakthroughs have resulted in successful test designs being synthesized even at 7 nm [48]. Many technology barriers were perceived along this path and scientists responded with innovations to circumvent, surmount, and tunnel through these challenges. Some of these innovations include the use of Strained-Silicon for 90 nm CMOS [23], High-κ/Metal Gate for 32 nm CMOS [43], and the Tri-Gate technology for 22 nm CMOS [32]. However, post 22 nm CMOS gate length offered severe challenges to the designers due to pronounced transistor short-channel effects, threatening the decline of Denned's Scaling principles and confounding the International Technology Roadmap for Semiconductors (ITRS) [30]. New research directions were sought and FinFET technology was adopted due to its superior short channel effects, achieving a gate

© Springer International Publishing AG 2018
A.K. Das et al., *Reliable and Energy Efficient Streaming Multiprocessor Systems*,
Embedded Systems, https://doi.org/10.1007/978-3-319-69374-3_1

length as low as 7 nm in 2013 [26, 48]. Alongside technology scaling, transistor integration capacity doubled during this time, enabling semiconductor vendors to pack millions of logic gates per chip.

1.1.2   Trends in Microprocessor Design

Transistor scaling has also led to the evolution of microprocessor technology since the first microprocessor was introduced in 1971 by Intel. This microprocessor used 2300 transistors and was based on 4-bit computations. Shortly afterwards, Intel developed the 8-bit microprocessor family—8008 and 8080 in 1972 and 1974, respectively. Thereafter, the microprocessor technology advanced to 16-bit (Intel 8086 in 1978), 32-bit (Intel 80386 in 1985), and finally to the Pentium series starting from 1993. Although microprocessor technology was pioneered by Intel, the journey saw several other key players such as Apple, AMD, and ARM among others, entering into the microprocessor market. Each generation of processor grew smaller and faster. This growth was guided mainly by the observation of Gordon Moore (co-founder of Intel), also known as Moore's Law, that computer performance will double every 18 months. Moore's Law together with Dennard's scaling principles enabled building faster and sophisticated microprocessors.

Figure 1.1 shows the trend in microprocessor development (source: Intel [63]). Throughout the 1990s and early 2000, microprocessor performance was synonymous with its operating frequency; higher frequency implied higher performance. This notion of performance was soon challenged and expanded to consider other key aspects, such as temperature, power dissipation, and energy consumption. In perspective of these new metrics, performance was predicted to deviate signif-

Fig. 1.1 Trend in microprocessor development (source: Intel [63])

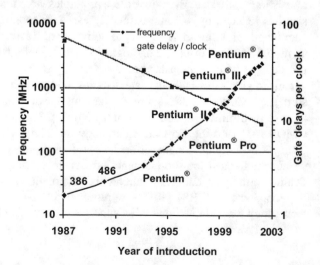

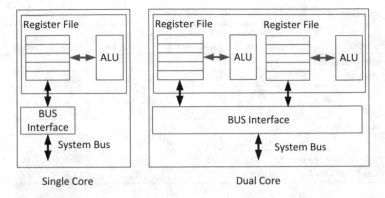

Fig. 1.2 Multi-core processor

icantly from the trend set by Moore's observation. This is because, with every new microprocessor generation, physical dimensions of chips decreased while the number of transistors per chip increased; the maximum clock frequency was thus bounded by a power envelop, crossing which could potentially burn the chip. Many techniques were sought to improve performance of microprocessor without further upscaling the frequency. One major breakthrough is that of multi-core processor— pioneered by research teams from IBM and Intel. A multi-core processor refers to a single computing unit with two or more microprocessor components (such as arithmetic and logic unit, ALU) such that multiple instructions can be read and executed in parallel. These parallel components are referred to as cores of a multi-core processor. Figure 1.2 shows a simplified representation of a single-core and a dual-core processor. The register file and the ALU of the single-core (shown in the red box) are replicated to form the dual-core processor. Other components such as the bus interface and the system bus are shared across the two cores.

Multi-core processors addressed thermal, power, and energy concerns. A processor with two cores running at a clock speed of f, can process the same number of instructions as that processed by a single-core processor running at $2f$ within the same time interval; yet the dual-core processor would still consume less energy. Furthermore, the power dissipation is distributed and so do thermal hotspots. This has motivated researchers over these years to integrate more cores in a single processor—2-cores (Intel Core 2 Duo, 2006), 4-cores (Intel Core i5, 2009), 6-cores (Intel Core i7, 2010), and 8-cores (Intel Xeon 2820, 2011). As the technology continued to scale further, the processor architecture started shifting from multi-core to many-core. Although there is no clear consensus among researchers to classify a processor as multi- or many-core, based on the number of comprising cores, we adopt one of the popular beliefs, classifying a processor with more than 8 cores as many-core processor. Table 1.1 reports the state-of-the-art multi- and many-core processors from some of the common microprocessor manufacturers.

Table 1.1 State-of-the-art multi- and many-core processors

Manufacturer	# Cores	Release year	Technology node (nm)	Max. power (W)	Max. frequency (GHz)
Multi-core processors					
IBM POWER7	8	2010	45	–	4.25
AMD Bulldozer	4	2011	32	125	3.9
Intel i7 Haswell	4	2013	22	84	3.5
Many-core processors					
Cell BE	9	2005	65	100	3.2
Oracle SPARC M6	12	2013	28	–	3.6
Intel Xeon Phi	60	2012	22	225	1.05

1.1.3 Trends in Multiprocessor Systems

As the semiconductor fabrication technology matured, there emerged a strong demand for single chip computing or System-on-Chip (SoC). An SoC is often defined as a single chip complex integrated circuit, incorporating the major processing elements of a complete end-product. Usually, an SoC comprises of a single full-fledged processor (acting as the master) to coordinate the operation of the other processing elements of the system. The SheevaPlug SoC [53] with one Kirkwood 6281 processor from Marvell Semiconductors is an example of a single-core SoC. As the processor technology shifted from single-core towards multi-core, SoC designers started integrating multi-core processors for faster coordination and processing. Intel's Cloverview [7] (based on dual-core Intel Atom) and Apple's A7 (based on dual-core ARM) are examples of multi-core SoCs. Finally, there are also commercially available SoCs featuring many-core processors. Examples are Sony's PlayStation 3 with IBM Cell processor with 9 cores (2006) and Texas Advanced Computing Center's Stampede SoC with 60-core Intel Xeon Phi processor (2013).

As the performance demand kept increasing, single processor SoCs soon became the performance bottleneck, even with up-scaled frequency and multi-core variants. Research took a new direction and thoughts were directed towards integrating multiple full-fledged processors (masters), to manage the application processes, alongside other hardware subsystems. Such platforms are commonly referred to as multiprocessor systems-on-chip (MPSoCs). The Lucent's Daytona chip [1], introduced in 2000, is the first known MPSoC, integrating multiple homogeneous processors. Figure 1.3 shows the Daytona MPSoC architecture. Following this breakthrough, there has been a significant drive towards MPSoC development especially in the early half of 2000. Examples included Nexperia [20] from Philips Semiconductor, OMAP [8] from Texas Instruments and Nomadik [3] from STMicroelectronics. Since then, the MPSoC technology has matured to a great extent, growing in complexity and size. Some of the current day MPSoCs are Uni-Phier [41] from Panasonic Semiconductor, Platform 2012 aka STHORM [40] from STMicroelectronics, and Tilera Gx8072 [58] from Tilera Corporation.

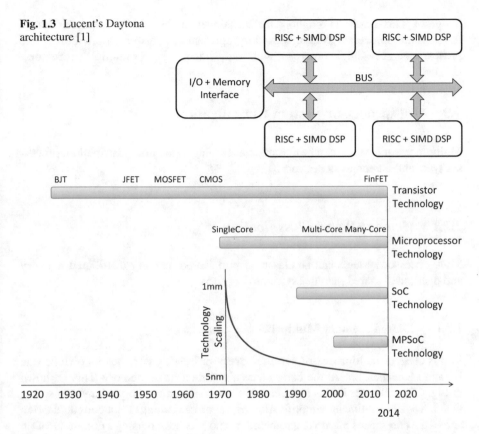

Fig. 1.3 Lucent's Daytona architecture [1]

Fig. 1.4 Summary of the technology trend

Developments in MPSoC have also triggered developments in the communication infrastructure, which interconnects the processing elements of these SoCs. Traditionally, MPSoCs integrated a shared bus, such as QuickPath Interconnect (QPI) from Intel, STBus from STMicrolelectronics, and the AMBA from ARM, to communicate among processing elements. However, the bottleneck soon shifted from computation capability to large communication delays, severely threatening the integration of additional processing elements. To circumvent this communication problem, a group of researchers from the Stanford University pioneered the implementation of networking concepts for data communication in multiprocessor system in 2002. This concept is known as network-on-chip (NoC) [4]. Since then, several commercial NoCs are developed across different research teams. Examples include Æthereal [24] and Hermes [39].

Summary of Trends Figure 1.4 shows the summary of the four technology trends—transistor technology, microprocessor technology, SoC technology and MPSoC technology. Shown in the same figure is the technology scaling following Dennards' Principles, starting with a gate length of 1 mm in 1974 to 5 nm in 2014.

Scope of This Book Throughout the remainder of this chapter (and the book), the term multiprocessor system is used to represent systems-on-chip with multiple processing cores—single processor with multi-/many-cores or multiple processors.

1.2   Multiprocessor System Classification

Multiprocessor systems can be classified according to memory distribution, processor type, and interconnect network.

1.2.1   Memory-Based Classification

Multiprocessor systems can be classified into shared memory distributed memory and distributed shared memory systems (Fig. 1.5).

1.2.1.1   Shared Memory Multiprocessor Systems

In this class of multiprocessor systems, every processing core has its own private L1 and L2 caches, but all the cores share a common main memory. This is shown in Fig. 1.5a. All cores in this architecture share a unique memory addressing space, which leads to significant simplification of the programming. Data communication across the cores takes place via the shared memory, accessed using a communication medium, typically a shared bus. Until recently, this was the dominating memory organization for multiprocessor systems with few cores (typically 2–4). However, as the gap between processor speed and the memory speed increases, and as more interacting parallel tasks are mapped to these cores, the memory bandwidth becomes a bottleneck for this class of systems.

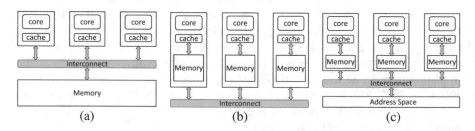

Fig. 1.5 Memory-based multiprocessor classification. (**a**) Shared memory. (**b**) Distributed memory. (**c**) Distributed shared memory

1.2.1.2 Distributed Memory Multiprocessor Systems

In this class of multiprocessor systems, every processing core has its own L1 and L2 caches and a private memory space. This is shown in Fig. 1.5b. This class of multiprocessor systems allows integration of any number of cores and offers a distinctive advantage over shared memory architecture in terms of scalability. However, programming is complex as compared to the shared memory counterpart. In this architecture, computational tasks mapped on a processing core can operate only on the local data; if remote data is required, the core communicates with other cores. Data communication takes place using a message passing interface through the interconnection network. One of the key characteristics of distributed memory architecture is its non-uniform memory access (NUMA) time. This is dependent on the interconnect network topology, such as Mesh, Tree, Star–Torus.

1.2.1.3 Distributed Shared Memory Multiprocessor Systems

Distributed shared memory architecture implements the shared memory abstraction on a multiprocessor system. Memories are physically separate and attached to individual core; however, the memory address space is shared across processing cores. This architecture combines the scalability of distributed architecture with the convenience of shared-memory programming. This architecture provides a virtual address space, shared among the cores. This is shown in Fig. 1.5c.

1.2.2 Processor-Based Classification

Multiprocessor systems can be classified into homogeneous and heterogeneous systems, based on the type of processing cores.

1.2.2.1 Homogeneous Multiprocessor Systems

A homogeneous multiprocessor system is a class of SoC where processing cores are of the same type. Examples of homogeneous multiprocessor systems are Philips Wasabi [57] and Lucent Daytona [1]. Homogeneous multiprocessor systems has low design replication effort, leading to high scalability and faster time-to-market.

1.2.2.2 Heterogeneous Multiprocessor Systems

Homogeneous architectures are often associated with high area and power requirements [34]; the current trend is, therefore, to integrate heterogeneous processing cores on the same SoC. Such architectures are referred to as heterogeneous

Table 1.2 Summary of homogeneous and heterogeneous architectures

	Homogeneous	Heterogeneous
Advantages	Less replication effort, highly scalable, faster time-to-market	Application specific, high computation efficiency, low power consumption
Limitations	Moderate computation efficiency, high power consumption	Less flexible, less scalable
Compatibility	Data parallelism, shared memory architecture, static and dynamic task mapping	Task parallelism, message passing interface, static task mapping
Examples	Lucent Daytona [1], Philip Wasabi [57]	Texas Instrument OMAP [8], Samsung Exynos 5 [49]

multiprocessor systems. This class of systems includes heterogeneity within and across processors. Texas Instrument OMAP [8] is an example of a heterogeneous multiprocessor system that integrates a traditional uni-core processor and a digital signal processor (DSP) core. Here the heterogeneity is across the different processors. On the other hand, Samsung Exynos 5 [49] is an example of an SoC that integrates a processor with heterogeneous cores (quad-core ARM big.LITTLE architecture [25] i.e., four ARM Cortex A7 cores and four Cortex A15 cores). One of the major limitations of heterogeneous multiprocessor systems is the overhead of integration. Table 1.2 reports the summary of these architectures.

An emerging trend in multiprocessor design is to integrate reconfigurable logic such as field programmable gate arrays (FPGAs) alongside homogeneous or heterogeneous cores [59]. This class of systems also belong to the heterogeneous category and are referred in literature as reconfigurable multiprocessor systems. These systems allow implementation of custom instructions [6] or custom logic [31] specific to an application. Examples include Xilinx Zynq [50].

Scope of This Book This book focuses on multiprocessor systems with distributed memory, using message passing interface for data communication between the processing cores. Both homogeneous and heterogeneous architectures (including the reconfigurable systems) are studied in this work.

1.3   Multiprocessor System Design Flow

A key concept in system design is the *orthogonalization of concerns* [33], i.e., the process of separating different aspects of the design in order to explore design spaces of individual aspects more efficiently. Figure 1.6 shows a typical multiprocessor design flow. The first stage of the design flow is the analysis stage. Input to this stage are a set of applications targeted for the platform and the platform's architectural specifications. Applications are represented as task graphs with nodes representing computation tasks of the application and edges representing dependency between

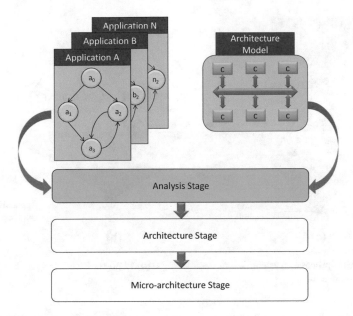

Fig. 1.6   Multiprocessor design flow

these tasks as shown in the figure. Application modeling is discussed further in Chap. 3. The system architecture is also represented as a connected graph, with nodes representing processing cores of the platform and edges representing the communication medium. Analysis stage of the design flow considers both, hardware and software aspects. An example hardware aspect is the number of processing cores needed for the platform to guarantee performances of the targeted applications while satisfying design budgets specified in terms of area, power, and reliability, among others. An example software aspect is the resource allocation, i.e., mapping of different tasks of each application on the processing cores.

The architecture stage determines architectural details of the resources selected in the analysis stage. For a general purpose processor, example architectural details include determining the instruction set. Architectural details for memory organization include size of the main memory for shared memory architecture or size of the private memories for the distributed memory architecture. Finally, the micro-architecture stage deals with the hardware of different components. The output of this stage is the register transfer logic (RTL) representation of the multiprocessor system, which can be synthesized on FPGA or ASIC.

Scope of This Book   The scope of this book is limited to the analysis stage of the design flow, which can be further classified into two types—platform-based design and hardware–software co-design.

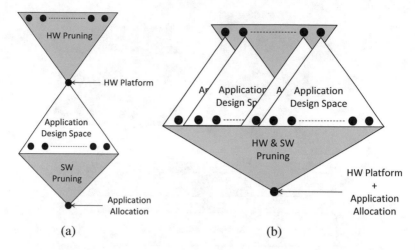

Fig. 1.7 Multiprocessor system design methodology. (**a**) Platform-based design. (**b**) Hardware–software co-design

1.3.1 Platform-Based Design

A platform-based design [44, 45] separates the hardware–software exploration process into two sequential processes—hardware exploration followed by the software exploration. This is shown in Fig. 1.7a. The hardware exploration is typically guided by design constraints such as chip area, power and reliability targets together with non-recurring engineering (NRE) costs. Once the target hardware platform is fixed, an application-level design space exploration is performed to allocate available resources of the hardware. The objective is to optimize metrics such as performance, energy, and reliability. Finally the application space is pruned to obtain resource allocation for every target application of the platform.

1.3.2 Hardware–Software Co-design

The hardware–software co-design [60, 61] performs hardware and software pruning steps simultaneously to obtain the target platform and targeted application's allocation on the platform. This is shown in Fig. 1.7b. The NRE design cost is usually considered as a constraint or is integrated with other optimization objectives such as performance, energy, and reliability. In its simplest form, the approach starts with a set of hardware design alternatives, each satisfying the NRE economics; the software goal is to explore the design space of allocating the resources for every design alternatives to optimize the required objectives. The hardware–software co-design achieves better optimization objectives than platform-based designs, at the price of higher exploration time and larger time-to-market.

1.4 Design Challenges for Multiprocessor Systems

Multiprocessor systems are widely used in embedded devices which are constrained in terms of battery life and thermal safety. This section highlights key concerns of modern multiprocessor systems, from embedded device perspective.

1.4.1 Energy Concern for Multiprocessor Systems

Traditionally, multiprocessor system designers have focused on improving performance. With increasing use of multiprocessor systems in embedded devices, energy optimization is becoming a primary objective. This is because by reducing the energy, it is possible to prolong the battery life embedded devices. Energy reduction can be achieved by (1) scaling the operating voltage and frequencies of the cores and (2) reducing data communication between dependent tasks.

Dynamic power consumption of a circuit can be expressed as $P = \alpha \cdot F \cdot C \cdot V^2$, where α is the switching factor, F is the frequency, C is the capacitance, and V is the voltage. Reducing both the voltage and the frequency by half reduces the total power dissipation by 87.5%. On the other hand, computation time of a job is inversely proportional to the frequency of operation and therefore, reducing the frequency by half increases the computation time two fold. The overall effect is that energy dissipation, measured as the product of power dissipation and computation time, reduces by 75% when both voltage and frequency are scaled by half. The process of scaling down voltage and frequency of operation dynamically during application execution is termed *dynamic voltage and frequency scaling* (DVFS) [19, 52, 55].

When dependent tasks of an application are mapped to different cores, energy is consumed to transfer data through the interconnect network. This energy is referred to as the communication energy, which can be as high as up to 40% of the total energy consumption for some applications [27]. One technique to mitigate this is to allocate communicating tasks on the same core. However, this impacts performance as tasks compete for shared compute and storage resources. To demonstrate this, a simple example is provided in Fig. 1.8. There are three tasks with their execution times (in ms) labeled as numbers inside the circle. Tasks B and C can execute after

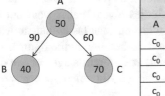

Allocation			Data Communication	Completion Time
A	B	C		
c_0	c_0	c_0	0	160
c_0	c_0	c_1	60	120
c_0	c_1	c_0	90	120
c_0	c_1	c_1	150	160

Fig. 1.8 Demonstration of the importance of communication energy

task A completes. The size of the data (in KB) required by tasks B and C from the output of task A is marked in the figure on arrows from A to B and A to C, respectively. The table shows different allocation of these tasks on a two-core system (identified as c_0 and c_1). Data communicated on the interconnect and completion time of the application are shown in columns 4 and 5, respectively. As can be clearly seen from this table, task allocations have impact on communication energy and computation time, requiring a trade-off analysis [54].

1.4.2   Reliability Concern for Multiprocessor Systems

Multiprocessor systems are expected to perform error-free operations. In this respect, an error of a system is defined as a malfunction, i.e., deviation from an expected behavior of the system. Errors are caused by faults. However, not every system fault leads to erroneous behavior. This book considers malignant faults, i.e., faults that are manifested as errors in the system. Such faults can be classified into three categories—transient, intermittent, and permanent. Transient faults are point failures, i.e., these faults occur once and then disappear. Primary causes for these faults are alpha or neutron emissions.[1] Several techniques have been proposed in literature for transient fault-tolerance. Examples are the use of error correction codes [22], checkpointing [51], rollback-recovery [46], and duplication with comparison [5], among others. Intermittent faults are a class of hardware defects, occurring frequently but irregularly over a period of time, due to process, voltage, and temperature (PVT) variations. There are techniques that optimize for intermittent faults by analyzing the steady-state availability [13]. Finally, permanent faults are damages to a circuit caused by such phenomena like electromigration, dielectric breakdowns, thermal fatigues etc. These faults, are caused during manufacturing or during a product's lifetime due to component wear-outs.

Permanent fault rate in an integrated circuit is described by the bathtub curve shown in Fig. 1.9. A high fault-rate is observed post manufacturing process due to non-ideal yield (<100%) of the technology node. Most of these faulty components are identified and separated (for further analysis) during testing phase. After this phase, probability of a hardware fault decreases. This time duration is known as the *infant mortality* period. This phase is followed by a period of constant failure rates, often referred as *useful life*. The last phase is known as the *wear-out* phase and is characterized by increasing fault rate. Recent studies reveal that if wear-out is not addressed early during useful life of a device, circuits can age faster than anticipated with the wear-out phase settling earlier in life (red dashed line in the figure).

Reliability optimization can also be performed at different stages of the design flow, with techniques adopted at each stage being orthogonal to one another.

---

[1]Transient faults in the look-up table of FPGA manifest as permanent faults until reprogrammed.

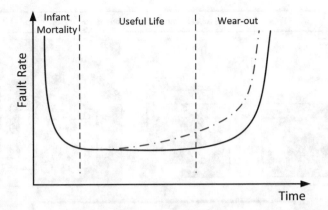

Fig. 1.9 Bathtub curve for permanent faults

Some examples of micro-architectural stage optimization techniques include the use of adaptive body biasing technique [36], use of 22 nm tri-gate transistor [47], and device-geometry aware design rule [37]. A summary of micro-architecture stage techniques is provided in [29]. Architecture-level techniques deal with processor instruction adaptations and scheduling. Examples include intelligent instruction scheduling [62], exploiting instruction timing slacks [42], and the compiler-directed register assignment [2]. Finally, reliability optimization at the analysis stage involves intelligent application mapping and scheduling [28].

1.5   Reliable and Energy Efficient Multiprocessor Design

This work solves the reliability and energy concerns for embedded multiprocessor systems, during the platform design stage (design-time) and also during in-field operation (run-time) of the final platform.

1.5.1   Design-Time Methodology

Design-time methodology of this work is highlighted in Fig. 1.10. In order to analyze temperature-related wear-outs of different mapping alternatives, a fast and accurate temperature model is developed. This model is characterized using temperature data obtained from industry-standard *HotSpot* [56] tool using the floorplan information of a given multiprocessor system. This thermal model is used in the platform-based design methodology along with target application and architecture models. Applications are represented as Synchronous Data Flow Graphs (SDFGs) [38] that allow modeling cyclic dependency, multi-input tasks,

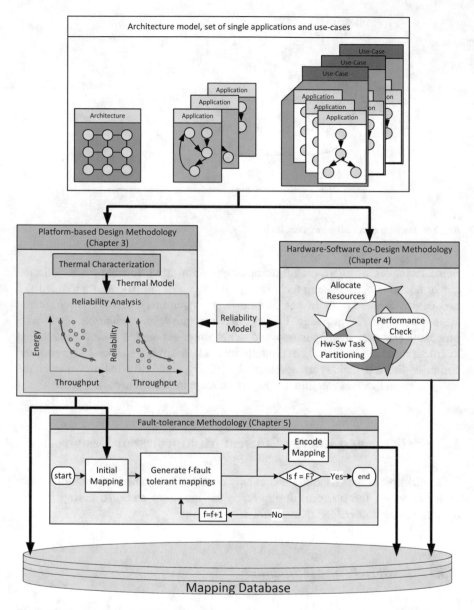

Fig. 1.10 Design-time methodology

multi-rate tasks, and pipelined execution. Multiprocessor systems are often designed for multiple applications, many of which are enabled concurrently (use-case). Formally, a use-case is defined as a collection of multiple applications that are active simultaneously on a multiprocessor system [35]. To allow optimization for concurrent applications, a set of use-cases is also considered in the design methodology. The target multiprocessor platform is represented as a directed graph with nodes representing processing cores and edges representing physical links between cores. The platform-based design methodology analyzes every application and use-case to determine mapping of computation tasks on the target platform such that temperature-related wear-outs and energy consumption are jointly minimized.

For emerging reconfigurable multiprocessor system such as the Zync, a hardware–software task partition flow is proposed, which decides computation tasks to be executed on the processing cores (ARM cores) and those to be implemented on the FPGA fabric. Based on this flow, a hardware–software co-design methodology is proposed to design reconfigurable multiprocessor systems. A set of applications and use-cases represented as SDFGs are used to determine the minimum number of processing cores and size of the reconfigurable area required for such systems to maximize lifetime reliability, while satisfying area and energy budgets.

The proposed platform-based design and the hardware–software co-design are proactive fault-tolerance approaches. A second aspect of the analysis phase of the design methodology is to ensure graceful performance degradation in the presence of faults. This methodology forms part of the reactive fault-tolerance approach, which is to deal with resource management post fault occurrences. To achieve this, analysis is performed to determine minimum energy mappings with least performance degradation for every fault-scenario. This analysis is performed for every target application for the system. Finally, to ensure that the performance requirement is satisfied using the pre-analyzed mappings at run-time, a self-timed execution-based scheduling technique is proposed for multiprocessor systems.

1.5.2   Run-Time Methodology

Figure 1.11 shows a high-level view of a typical multiprocessor system. At the top is the application layer, composed of a set of user-defined applications. At the bottom is the hardware layer, which consists of processing cores. In between these two layers is the operating system or the system software layer, which coordinates application execution on the hardware. The operating system is responsible to assign computation tasks on the hardware cores. If the application to be executed on the hardware is already analyzed at design-time, the operating system performs task allocation according to the pre-computed mapping. On the other hand, if a new application (not analyzed beforehand) needs to be executed on the hardware, the operating system needs to determine the correct mapping such that temperature-related wear-out can be minimized. To achieve this, a machine learning-based run-time approach is developed in this work. The machine learning framework is integrated as a part of the operating system and interacts with the application layer, on the one hand, to determine performance requirements and with the hardware, on the other hand, to collect thermal statistics. Based on these information, the operating system adjusts voltage-frequency settings and application threads to reduce thermal emergencies.

Fig. 1.11 Run-time
methodology

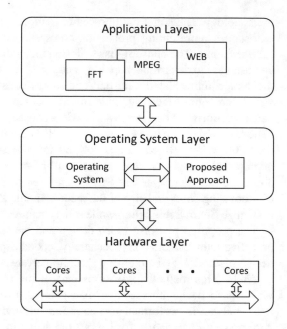

1.6 Key Highlights of This Book

Following are key contributions of this book to the multiprocessor community:

1. *A fast and accurate thermal model* to estimate temperature of all cores of a
   multiprocessor system from a given floorplan. This model incorporates both
   temporal and spatial dependencies. Predicted temperature from this model is
   shown to be within 4% of state-of-the-art accurate thermal model.[2] This thermal
   model is published in [17, 18].
2. *A gradient-based fast heuristic* to jointly optimize lifetime reliability and energy
   consumption of a given multiprocessor system with applications represented as
   SDFGs. The proposed methodology increases system lifetime by an average 47%
   and minimizes energy consumption by 24% with respect to existing techniques.
   This work is published in [11].
3. *A hardware–software task partitioning framework* for reconfigurable multiprocessor system to decide on computation tasks to be executed on processing
   cores and those on the reconfigurable area. The objective is to improve the
   lifetime reliability of the platform while exploring the trade-off between lifetime
   reliability and the reliability considering transient faults with checkpointing-

---

[2]The state-of-the-art thermal models are too computation intensive to be included in the design
space exploration process.

based fault-tolerance. The approach is shown to improve lifetime reliability by an average 60% with less than 15% sacrifice of reliability considering transient faults. This work is published in [12].

4. *A hardware–software co-design methodology* to determine the minimum number of cores and the size of the reconfigurable area for a reconfigurable multiprocessor system such that the lifetime reliability is maximized while satisfying the area and energy budget. This methodology is shown to improve lifetime reliability by an average 65% for single applications and an average 70% for use-cases with 25% fewer cores and 20% less reconfigurable area as compared to the existing hardware–software co-design approaches.

5. *A design-time (off-line) multi-criteria optimization technique* for application mapping on multiprocessor systems to minimize the energy consumption for all processor fault-scenarios while providing graceful throughput degradation. The proposed technique is shown to minimize energy consumption by an average 22% as compared to existing techniques. Parts of this contribution are first published in [10] and [9], and later improved and published in [14] and [15], respectively.

6. *A scheduling technique based on self-timed execution* to minimize schedule storage and reconstruction overhead at run-time. The scheduling technique is shown to minimize schedule construction time by 95% and storage overhead by 92%. This work is published in [15].

7. *A reinforcement learning-based inter-and intra-application thermal management* to control peak temperature as well as thermal cycling using thread-to-core allocation (through CPU affinity) and dynamic frequency control (through CPU governors). Results demonstrate that the proposed approach minimizes average temperature, peak temperature, and thermal cycling, improving the mean time to failure by an average of $2\times$ for intra-application and $3\times$ for inter-application scenarios when compared to existing thermal management techniques. Furthermore, dynamic and static energy consumption are also reduced by an average 10% and 11%, respectively. This work is published in [16].

The remainder of this book is organized as follows. Chapter 2 introduces application, architecture, and reliability models used in the proposed approach. Chapter 3 provides an overview literature survey of the existing techniques for reliability and energy optimization in multiprocessor systems, identifying the gaps in these studies. Chapter 4 introduces reliability and energy-aware platform-based design methodologies for multiprocessor system. Hardware–software co-design methodology for reconfigurable multiprocessor system is introduced in Chap. 5. The reactive fault-tolerance methodology is discussed next in Chap. 6. The run-time thermal management is discussed in Chap. 7. Finally, Chap. 8 concludes this book with an overview of the future research directions.

References

1. B. Ackland, A. Anesko, D. Brinthaupt, S. Daubert, A. Kalavade, J. Knobloch, E. Micca, M. Moturi, C. Nicol, J. O'Neill, J. Othmer, E. Sackinger, K. Singh, J. Sweet, C. Terman, J. Williams, A single-chip, 1.6-billion, 16-b MAC/s multiprocessor DSP. IEEE J. Solid State Circuits **35**(3), 412–424 (2000)
2. F. Ahmed, M. Sabry, D. Atienza, L. Milor, Wearout-aware compiler-directed register assignment for embedded systems, in *Proceedings of the International Symposium on Quality Electronic Design (ISQED)* (2012), pp. 33–40
3. A. Artieri, V. Alto, R. Chesson, M. Hopkins, M. Rossi, Nomadik open multimedia platform for next-generation mobile devices, in *STMicroelectronics Technical Article TA305* (2003)
4. L. Benini, G. De Micheli, Networks on chips: a new SoC paradigm. IEEE Comput. **35**(1), 70–78 (2002)
5. C. Bolchini, A. Miele, Reliability-driven system-level synthesis for mixed-critical embedded systems. IEEE Trans. Comput. **62**(12), 2489–2502 (2013)
6. L. Chen, T. Mitra, Shared reconfigurable fabric for multi-core customization, in *Proceeding of the Annual Design Automation Conference (DAC)* (ACM, 2011), pp. 830–835
7. Cloverview Family of Intel SoC, Intel Corporation (2009), http://ark.intel.com/products/codename/53606/cloverview
8. P. Cumming, The TI OMAP platform approach to SoC, in *Winning the SOC Revolution* (2003)
9. A. Das, A. Kumar, Fault-aware task re-mapping for throughput constrained multimedia applications on NoC-based MPSoCs, in *Proceedings of the International Symposium on Rapid System Prototyping (RSP)* (IEEE, 2012), pp. 149–155
10. A. Das, A. Kumar, B. Veeravalli, Energy-aware communication and remapping of tasks for reliable multimedia multiprocessor systems, in *Proceedings of the International Conference on Parallel and Distributed Systems (ICPADS)* (IEEE Computer Society, 2012), pp. 564–571
11. A. Das, A. Kumar, B. Veeravalli, Reliability-driven task mapping for lifetime extension of networks-on-chip based multiprocessor systems, in *Proceedings of the Conference on Design, Automation and Test in Europe (DATE)* (European Design and Automation Association, 2013), pp. 689–694
12. A. Das, A. Kumar, B. Veeravalli, Aging-aware hardware-software task partitioning for reliable reconfigurable multiprocessor systems, in *Proceedings of the International Conference on Compilers, Architecturesand Synthesis for Embedded Systems (CASES)* (IEEE Press, 2013), pp. 1:1–1:10
13. A. Das, A. Kumar, B. Veeravalli, Communication and migration energy aware design space exploration for multicore systems with intermittent faults, in *Proceedings of the Conference on Design, Automation and Test in Europe (DATE)* (European Design and Automation Association, 2013), pp. 1631–1636
14. A. Das, A. Kumar, B. Veeravalli, Communication and migration energy aware task mapping for reliable multiprocessor systems. Elsevier Futur. Gener. Comput. Syst. **30**, 216–228 (2014)
15. A. Das, A. Kumar, B. Veeravalli, Energy-aware task mapping and scheduling for reliable embedded computing systems. ACM Trans. Embed. Comput. Syst. (TECS) **13**(2s), 72:1–72:27 (2014)
16. A. Das, R.A. Shafik, G.V. Merrett, B.M. Al-Hashimi, A. Kumar, B. Veeravalli, Reinforcement learning-based inter- and intra-application thermal optimization for lifetime improvement of multicore systems, in *Proceeding of the Annual Design Automation Conference (DAC)* (ACM, 2014)
17. A. Das, A. Kumar, B. Veeravalli, Temperature aware energy-reliability trade-offs for mapping of throughput-constrained applications on multimedia MPSoCs, in *Proceedings of the Conference on Design, Automation and Test in Europe (DATE)* (European Design and Automation Association, 2014)

18. A. Das, A. Kumar, B. Veeravalli, Reliability and energy-aware mapping and scheduling of multimedia applications on multiprocessor systems. IEEE Trans. Parallel Distrib. Syst. **27**(3), 869–884 (2016)
19. M. Damavandpeyma, S. Stuijk, T. Basten, M. Geilen, H. Corporaal, Throughput-constrained DVFS for scenario-aware dataflow graphs, in *Proceedings of the IEEE Symposium on Real-Time and Embedded Technology and Applications (RTAS)* (2013), pp. 175–184
20. J. De Oliveira, H. Van Antwerpen, The Philips Nexperia digital video platform, in *Winning the SoC Revolution* (2003), pp. 67–96
21. R. Dennard, F. Gaensslen, V. Rideout, E. Bassous, A. LeBlanc, Design of ion-implanted MOSFET's with very small physical dimensions. IEEE J. Solid State Circuits **9**(5), 256–268 (1974)
22. M. Fojtik, D. Fick, Y. Kim, N. Pinckney, D. Harris, D. Blaauw, D. Sylvester, Bubble razor: eliminating timing margins in an ARM Cortex-M3 processor in 45 nm CMOS using architecturally independent error detection and correction. IEEE J. Solid State Circuits **48**(1), 66–81 (2013)
23. T. Ghani, M. Armstrong, C. Auth, M. Bost, P. Charvat, G. Glass, T. Hoffmann, K. Johnson, C. Kenyon, J. Klaus, B. McIntyre, K. Mistry, A. Murthy, J. Sandford, M. Silberstein, S. Sivakumar, P. Smith, K. Zawadzki, S. Thompson, M. Bohr, A 90nm high volume manufacturing logic technology featuring novel 45nm gate length strained silicon CMOS transistors, in *IEEE International Electron Devices Meeting (IEDM)* (2003), pp. 11.6.1–11.6.3
24. K. Goossens, J. Dielissen, A. Radulescu, Æthereal network on chip: concepts, architectures, and implementations. IEEE Des. Test Comput. **22**(5), 414–421 (2005)
25. P. Greenhalgh, big.LITTLE processing with ARM Cortex-A15 & Cortex-A7, ARM White Paper (2011), http://www.arm.com/products/processors/technologies/biglittleprocessing.php
26. S. Gupta, V. Moroz, L. Smith, Q. Lu, K. Saraswat, A group IV solution for 7 nm FinFET CMOS: stress engineering using Si, Ge and Sn, in *IEEE International Electron Devices Meeting (IEDM)* (2013), pp. 26.3.1–26.3.4
27. J. Hu, R. Marculescu, Energy-aware communication and task scheduling for network-on-chip architectures under real-time constraints, in *Proceedings of the Conference on Design, Automation and Test in Europe(DATE)* (IEEE Computer Society, 2004), p. 10234
28. L. Huang, F. Yuan, Q. Xu, Lifetime reliability-aware task allocation and scheduling for MPSoC platforms, in *Proceedings of the Conference on Design, Automation and Test in Europe (DATE)* (European Design and Automation Association, 2009), pp. 51–56
29. V. Huard, C. Parthasarathy, A. Bravaix, C. Guerin, E. Pion, CMOS device design-in reliability approach in advanced nodes, in *IEEE International Reliability Physics Symposium* (2009), pp. 624–633
30. International technology roadmap for semiconductors (ITRS), Semiconductor Industry Association (2008), http://www.itrs.net
31. L. Jiashu, A. Das, A. Kumar, A design flow for partially reconfigurable heterogeneous multi-processor platforms, in *Proceedings of the International Symposium on Rapid System Prototyping (RSP)* (IEEE, 2012), pp. 170–176
32. E. Karl, Y. Wang, Y.-G. Ng, Z. Guo, F. Hamzaoglu, U. Bhattacharya, K. Zhang, K. Mistry, M. Bohr, A 4.6GHz 162Mb SRAM design in 22nm tri-gate CMOS technology with integrated active VMIN-enhancing assist circuitry, in *Proceedings of the IEEE International Solid-State Circuits Conference Digest of Technical Papers (ISSCC)* (2012), pp. 230–232
33. K. Keutzer, A. Newton, J. Rabaey, A. Sangiovanni-Vincentelli, System-level design: orthogonalization of concerns and platform-based design. IEEE Trans. Comput. Aided Des. Integr. Circuits Syst. (TCAD) **19**(12), 1523–1543 (2000)
34. R. Kumar, D.M. Tullsen, P. Ranganathan, N.P. Jouppi, K.I. Farkas, Single-ISA heterogeneous multi-core architectures for multithreaded workload performance, in *Proceedings of the Annual International Symposium on Computer Architecture (ISCA)* (IEEE Computer Society, 2004), pp. 64–75

35. A. Kumar, B. Mesman, H. Corporaal, Y. Ha, Iterative probabilistic performance prediction for multi-application multiprocessor systems. IEEE Trans. Comput. Aided Des. Integr. Circuits Syst. (TCAD) **29**(4), 538–551 (2010)
36. S. Kumar, C. Kim, S. Sapatnekar, Adaptive techniques for overcoming performance degradation due to aging in CMOS circuits. IEEE Trans. Very Large Scale Integr. Syst. (TVLSI) **19**(4), 603–614 (2011)
37. Y. Leblebici, Design considerations for CMOS digital circuits with improved hot-carrier reliability. IEEE J. Solid State Circuits **31**(7), 1014–1024 (1996)
38. E. Lee, D. Messerschmitt, Synchronous data flow. Proc. IEEE **75**(9), 1235–1245 (1987)
39. F. Moraes, N. Calazans, A. Mello, L. Möller, L. Ost, HERMES: an infrastructure for low area overhead packet-switching networks on chip. Integr. VLSI J. **38**(1), 69–93 (2004)
40. J. Mottin, M. Cartron, G. Urlini, The STHORM platform, in *Embedded Systems*, ed. by M. Torquati, K. Bertels, S. Karlsson, F. Pacull (Springer, New York, 2014), pp. 35–43
41. Y. Nishimichi, N. Higaki, M. Osaka, S. Horii, H. Yoshida, UniPhier: series development and SoC management, in *Proceedings of the Asia and South Pacific Design Automation Conference (ASP-DAC)*, ser. ASP-DAC '09 (IEEE Press, 2009), pp. 540–545
42. F. Oboril, F. Firouzi, S. Kiamehr, M. Tahoori, Reducing NBTI-induced processor wearout by exploiting the timing slack of instructions, in *Proceedings of the Conference on Hardware/Software Codesign and System Synthesis (CODES+ISSS)*, ser. CODES+ISSS '12 (ACM, 2012), pp. 443–452
43. P. Packan, S. Akbar, M. Armstrong, D. Bergstrom, M. Brazier, H. Deshpande, K. Dev, G. Ding, T. Ghani, O. Golonzka, W. Han, J. He, R. Heussner, R. James, J. Jopling, C. Kenyon, S.-H. Lee, M. Liu, S. Lodha, B. Mattis, A. Murthy, L. Neiberg, J. Neirynck, S. Pae, C. Parker, L. Pipes, J. Sebastian, J. Seiple, B. Sell, A. Sharma, S. Sivakumar, B. Song, A. St.Amour, K. Tone, T. Troeger, C. Weber, K. Zhang, Y. Luo, S. Natarajan, High performance 32nm logic technology featuring 2nd generation high-k + metal gate transistors, in *IEEE International Electron Devices Meeting (IEDM)* (2009), pp. 1–4
44. A. Pinto, A. Bonivento, A.L. Sangiovanni-Vincentelli, R. Passerone, M. Sgroi, System level design paradigms: platform-based design and communication synthesis, in *Proceeding of the Annual Design Automation Conference (DAC)* (ACM, 2004), pp. 537–563
45. K. Popovici, X. Guerin, F. Rousseau, P.S. Paolucci, A.A. Jerraya, Platform-based software design flow for heterogeneous MPSoC. ACM Trans. Embed. Comput. Syst. (TECS) **7**(4), 39:1–39:23 (2008)
46. M. Prvulovic, Z. Zhang, J. Torrellas, ReVive: cost-effective architectural support for rollback recovery in shared-memory multiprocessors, in *Proceedings of the Annual International Symposium on Computer Architecture (ISCA)* (IEEE Computer Society, 2002), pp. 111–122
47. S. Ramey, A. Ashutosh, C. Auth, J. Clifford, M. Hattendorf, J. Hicks, R. James, A. Rahman, V. Sharma, A. St.Amour, C. Wiegand, Intrinsic transistor reliability improvements from 22nm tri-gate technology, in *IEEE International Reliability Physics Symposium (IRPS)* (2013), pp. 4C.5.1–4C.5.5
48. J. Ryckaert, Scaling beyond 7nm: design-technology co-optimization at the rescue, in *Proceedings of the 2016 on International Symposium on Physical Design* (ACM, 2016), p. 89
49. Samsung Exynos 5 Octa, Samsung Electronics (2014), www.samsung.com/exynos
50. M. Santarini, Zynq-7000 EPP sets stage for new era of innovations. Xcell J. **75**, 8–13 (2011)
51. P.K. Saraswat, P. Pop, J. Madsen, Task mapping and bandwidth reservation for mixed hard/soft fault-tolerant embedded systems, in *Proceedings of the IEEE Symposium on Real-Time and Embedded Technology and Applications (RTAS)* (IEEE Computer Society, 2010), pp. 89–98
52. M.T. Schmitz, B.M. Al-Hashimi, P. Eles, Cosynthesis of energy-efficient multimode embedded systems with consideration of mode-execution probabilities. IEEE Trans. Comput. Aided Des. Integr. Circuits Syst. (TCAD) **24**(2), 153–169 (2005)
53. SheevaPlug SoC, Marvell Semiconductor (2009), http://www.marvell.com/solutions/plug-computers/
54. A.K. Singh, T. Srikanthan, A. Kumar, W. Jigang, Communication-aware heuristics for run-time task mapping on noc-based mpsoc platforms. J. Syst. Archit. **56**(7), 242–255 (2010)

55. A.K. Singh, A. Das, A. Kumar, Energy optimization by exploiting execution slacks in streaming applications on multiprocessor systems, in *Proceeding of the Annual Design Automation Conference (DAC)* (ACM, 2013), pp. 115:1–115:7

56. K. Skadron, M.R. Stan, K. Sankaranarayanan, W. Huang, S. Velusamy, D. Tarjan, Temperature-aware microarchitecture: modeling and implementation, ACM Trans. Archit. Code Optim. (TACO) **1**(1), 94–125 (2004)

57. P. Stravers, J. Hoogerbrugge, Homogeneous multiprocessing and the future of silicon design paradigms, in *Proceedings of the International Symposium on VLSI Technology, Systems, and Applications* (2001), pp. 184–187

58. TILE-Gx8072 processor specification. Tilera Corporation (2014), http://www.tilera.com/sites/default/files/images/products/TILE-Gx8072_PB041-03_WEB.pdf

59. M.A. Watkins, D.H. Albonesi, ReMAP: a reconfigurable heterogeneous multicore architecture, in *Proceedings of the IEEE/ACM International Symposium on Microarchitecture (MICRO)* (IEEE Computer Society, 2010), pp. 497–508

60. W. Wolf, Hardware-software co-design of embedded systems. Proc. IEEE **82**(7), 967–989 (1994)

61. J. Teich, Hardware/software codesign: the past, the present, and predicting the future. Proc. IEEE, **100**(Special Centennial Issue), 1411–1430 (2012)

62. A. Tiwari, J. Torrellas, Facelift: hiding and slowing down aging in multicores, in *Proceedings of the IEEE/ACM International Symposium on Microarchitecture (MICRO)* (IEEE Computer Society, 2008), pp. 129–140

63. R. Yung, S. Rusu, K. Shoemaker, Future trend of microprocessor design, in *Proceedings of the European Solid-State Circuits Conference (ESSCIRC)* (IEEE, 2002), pp. 43–46

Chapter 2
Operational Semantics of Application and Reliability Model

2.1 Application Model as Synchronous Data Flow Graphs

Synchronous Data Flow Graphs (SDFGs, see [12]) are often used for modeling modern DSP applications [17] and for designing concurrent multimedia applications implemented on multiprocessor systems. Both pipelined streaming and cyclic dependencies between tasks can be easily modeled in SDFGs. SDFGs allow analysis of a system in terms of throughput and other performance properties, e.g., latency and buffer requirements [18]. Nodes of an SDFG are called *actors*; they represent functions that are computed by reading *tokens* (data items) from their input ports and writing results of computation as tokens on output ports. The number of tokens produced or consumed in one execution of an actor is called *rate*, and remains constant. Rates are visualized as port annotations. Actor execution is also called *firing*, and requires a fixed amount of time, denoted with a number in the actors. Edges in the graph, called *channels*, represent dependencies among different actors.

Figure 2.1 shows an example of an SDFG. There are three actors in this graph. In the example, a_0 has an input rate of 1 and output rate of 2. An actor is called *ready* when it has sufficient input tokens on all its input edges and sufficient buffer space on all its output channels; an actor can only fire when it is ready. The edges may also contain *initial tokens*, indicated by bullets on the edges, as seen on the edge from actor a_2 to a_0 in Fig. 2.1. Formally, an SDFG is a directed graph $\mathscr{G}_{app} = (A, C)$ consisting of a finite set A of actors and a finite set C of channels.

One of the most interesting properties of SDFGs is throughput, which is defined as the inverse of the long-term period, i.e., the average time needed for one iteration of the application. An iteration is defined as the minimum non-zero execution such that the original state of the graph is obtained. This is the performance metric used throughout the rest of this book. The following properties of an SDFG are defined.

© Springer International Publishing AG 2018   23
A.K. Das et al., *Reliable and Energy Efficient Streaming Multiprocessor Systems*,
Embedded Systems, https://doi.org/10.1007/978-3-319-69374-3_2

Fig. 2.1 Application SDFG

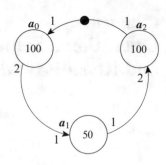

Definition 1 (Repetition Vector) Repetition Vector *RV* of an SDFG, $\mathscr{G}_{app}$ = (A, C) is defined as the vector specifying the number of times actors in A are executed for one iteration of SDFG G_{app}. In Fig. 2.1, $RV[a_0 \; a_1 \; a_2] = [1 \; 2 \; 1]$.

Definition 2 (Application Period) Application Period *Per(A)* is defined as the time SDFG, $\mathscr{G}_{app}$ = (A, C) takes to complete one iteration on an average.

Period of an SDFG can be computed by analyzing the maximum cycle mean (MCM) of an equivalent homogeneous SDFG (HSDFG). The period thus computed gives the minimum period possible with infinite hardware resources, e.g. buffer space. If worst-case execution time estimates of each actor are used, the performance at run-time is guaranteed to meet the analyzed throughput. Self-timed strategy is widely used for scheduling SDFGs on multiprocessor systems. In this technique, exact firing of an actor on a core is determined at design-time using worst-case actor execution-time. The timing information is then discarded retaining the assignment and ordering of actors on each core. At run-time, actors are fired in the same order as determined from design-time. Thus, unlike fully-static schedules, a self-timed schedule is robust in capturing the dynamism in actor execution time. The self-timed execution of an SDFG consists of a transient phase followed by a periodic or steady-state phase [8].

2.2   Wear-Out Related Reliability Model

This section reviews the reliability model of multiprocessor systems. We use a bottom-up approach, starting with the device-level models. These models are integrated to processor (core)-level models, which are integrated to generate multiprocessor system-level model.

2.2.1   Device-Level Reliability Modeling

Following are the temperature-dependent dominant wear-out failures in semiconductor devices.

2.2.1.1 Electromigration

Electromigration refers to movement of metal atoms from interconnect wires and vias due to the flow of current, temperature gradient, and electric diffusion, causing open and short circuits in interconnects. The mean time to failure (MTTF) due to electromigration is given by the following equation [3, 4].

$$\text{MTTF}_{\text{EM}} = \frac{A_{\text{EM}}}{J^n} \exp\left(\frac{E_{a_{\text{EM}}}}{KT}\right) \tag{2.1}$$

where A_{EM} is a material-dependent constant, J is the current density, n is empirically determined constant with a typical value of 2 for stress related failures, $E_{a_{\text{EM}}}$ is the activation energy of electromigration, K is the Boltzmann's constant, and T is the temperature. The current density J is determined as

$$J = \frac{f \cdot C \cdot V_{\text{dd}} \cdot P_t}{W \cdot H} \tag{2.2}$$

where C is the parasitic capacitance, V_{dd} is the supply voltage, f is the clock frequency, P_t is the probability of line toggling in a clock cycle, W is the width of the metal line, and H is the thickness of the metal line.

2.2.1.2 Negative Bias Temperature Instability

Negative bias temperature instability refers to shift in the threshold voltage and saturation current in p-channel MOS (PMOS) devices after an extended period of stress caused by applying negative voltage across gate to channel region [1, 23]. MTTF due to negative bias temperature instability is given by the following equation [2]

$$\text{MTTF}_{\text{NBTI}} = \frac{A_{\text{NBTI}}}{(V_{\text{GS}})^\gamma} \exp\left(\frac{E_{a_{\text{NBTI}}}}{KT}\right) \tag{2.3}$$

where A_{NBTI} is a constant dependent on the fabrication process, V_{GS} is the gate voltage, γ is the voltage acceleration factor, and $E_{a_{\text{NBTI}}}$ is the activation energy of negative bias temperature instability.

2.2.1.3 Hot Carrier Injection

There are three types of carrier injection—channel hot electron injection, drain avalanche hot carrier injection, and secondary generated hot electron injection. Channel hot electron injection refers to the escape of electrons from the channel causing a degradation of Si–SiO$_2$ interface [19, 20]. Drain avalanche hot carrier injection refers to gate oxide degradation due to electron and hole gate currents caused due to the impact ionization [19]. Finally, secondary generated hot electron injection refers to injection of minority carriers due to secondary impact ionization [19]. MTTF due to hot carrier injection is given by

$$\text{MTTF}_{\text{HCI}} = A_{\text{HCI}} \exp\left(\frac{\theta}{V_{\text{DS}}}\right) \tag{2.4}$$

where A_{HCI} and θ are empirically determined constants and V_{DS} is the drain to source voltage.

2.2.1.4   Time-Dependent Dielectric Breakdown

Time-dependent dielectric breakdown refers to degradation of the SiO_2 insulating layer between gate and conducting channel of a MOS device. Applied voltage and tunneling electrons create defects in the volume of oxide film, which accumulates over time triggering a sudden loss of dielectric properties. The exact physical mechanism behind the degradation is still an open question. MTTF due to time-dependent dielectric breakdown is given by [3]

$$\text{MTTF}_{\text{TDDB}} = A_{\text{TDDB}} \cdot A_G \cdot \left(\frac{1}{V_{\text{GS}}}\right)^{\alpha - \beta T} \exp\left(\frac{X}{T} + \frac{Y}{T^2}\right) \tag{2.5}$$

where α, β, X, and Y are fitting parameters, A_G is the surface area of gate oxide, and A_{TDDB} is an empirically determined constant.

2.2.1.5   Stress Migration

Stress migration refers to motion of atoms in metal wires due to mechanical stress caused by mismatch of temperature between metal and dielectric material. MTTF due to stress migration is given by

$$\text{MTTF}_{\text{SM}} = A_{\text{SM}} |T_0 - T|^{-n} \exp\left(\frac{E_{a_{\text{SM}}}}{KT}\right) \tag{2.6}$$

where A_{SM} is a material-dependent constant, T_0 is the metal deposition temperature, and $E_{a_{\text{SM}}}$ is the activation energy of stress migration.

2.2.1.6   Thermal Cycling

Thermal cycling refers to wear-out caused by thermal stress due to a mismatched coefficient of thermal expansion of adjacent material layers. Thermal cycling related MTTF is computed in three steps.

1. Calculating thermal cycles from a thermal profile using Downing simple rainflow counting algorithm [7].
2. Calculating, from each thermal cycle, the number of cycles to failure using Coffin-Manson's rule [14].

$$N_{TC}(i) = A_{TC} \left(\delta T_i - T_{Th}\right)^{-b} e^{\frac{E_{a_{TC}}}{KT_{max}(i)}} \tag{2.7}$$

where $N_{TC}(i)$ is the number of cycles to failure due to ith thermal cycle, A_{TC} is an empirically determined constant, δT_i is the amplitude of the ith thermal cycle, T_{Th} is the temperature at which elastic deformation begins, b is the Coffin-Manson exponent constant, $E_{a_{TC}}$ is the activation energy of thermal cycling, and $T_{max}(i)$ is the maximum temperature in the ith thermal cycle.
3. Calculating the MTTF using Miner's rule [15].

$$\text{MTTF} = \frac{N_{TC} \sum_{i=1}^{m} t_i}{m} \tag{2.8}$$

where t_i is the time for the ith thermal cycle, m is the number of thermal cycles obtained in step 1, and N_{TC} is the effective cycles to failure determined using

$$N_{TC} = \frac{m}{\sum_{i=1}^{m} \frac{1}{N_{TC}(i)}} \tag{2.9}$$

Combining Eqs. (2.7)–(2.9),

$$\text{MTTF} = \frac{A_{TC} \sum_{i=1}^{m} t_i}{\text{Thermal Stress}} \tag{2.10}$$

where "Thermal Stress" is an indication of the stress experienced due to the thermal cycling. This is obtained using the following equation.

$$\text{Thermal Stress} = \sum_{i=1}^{m} (\delta T_i - T_{Th})^b \times e^{\frac{-E_a}{KT_{max}(i)}} \tag{2.11}$$

2.2.2   Core-Level Reliability Modeling

Core-level reliability modeling involves combining device-level reliability models to estimate MTTF of a core of a multiprocessor system. Fault density at the device level is typically characterized by Weibull or Lognormal distribution. For example, time-dependent dielectric breakdown follows Weibull distribution and electromigration follows Lognormal distribution. The distributions for other wear-out mechanisms are not known with certainty. Reliability computation is demonstrated for these two types of distribution.

2.2.2.1   Weibull Distribution

The reliability of a core considering Weibull distribution is given by

$$R(t) = e^{-\left(\frac{t}{\eta}\right)^{\beta}} \tag{2.12}$$

where η is the scale parameter and β is the shape parameter. When temperature is not a constant, but varies over time, any time interval 0 to t can be split into n disjoint time intervals $\{[0, t_1), [t_1, t_2). \cdots [t_{n-1}, t_n)\}$ such that the temperature is constant within each interval. Let T_i be the temperature at the time interval $[t_i, t_{i+1})$. The scale parameter for the Weibull distribution for each interval is given by

$$\eta_i = \frac{\text{MTTF}_{\text{WO}}(T_i)}{\Gamma(1+\beta)} \tag{2.13}$$

where Γ is the gamma function and MTTF_{WO} is the MTTF with specific wear-out (WO) type under consideration, i.e.

$$\text{WO} = \begin{cases} \text{EM} & \text{for electromigration} \\ \text{NBTI} & \text{for negative bias temperature instability} \\ \cdots \end{cases} \tag{2.14}$$

The reliability is therefore given by [22]

$$R(t) = e^{-\left(\sum_{i=1}^{n} \frac{t_i - t_{i-1}}{\eta_i}\right)^{\beta}} \tag{2.15}$$

Impact of Process Variation   The expression for reliability is derived assuming a constant value for the shape parameter β. When process variation is considered, β is not constant, but is given by a Gaussian distribution function $\phi(\mu_g, \sigma_g)$, where μ_g is the mean and σ_g^2 is the variance. The reliability of a component (core) with N devices considering process variation is given by [6]

$$R(t) = e^{-N\left(\sum_{i=1}^{n} \frac{t_i - t_{i-1}}{\eta_i}\right)^{\mu_g - \frac{\sigma_g^2}{2} \ln\left(\sum_{i=1}^{n} \frac{t_i - t_{i-1}}{\eta_i}\right)}} \tag{2.16}$$

2.2.2.2   Lognormal Distribution

The reliability of a core considering Lognormal distribution is given by

$$R(t) = \frac{1}{2} - \frac{1}{2}\text{erf}\left(\frac{\ln(t) - \mu}{\sqrt{2\sigma^2}}\right) \tag{2.17}$$

where μ is the scale parameter, σ is the shape parameter, and erf is the error function. Considering a time distribution of temperature as before, the reliability is

$$R(t) = \frac{1}{2} - \frac{1}{2}\mathrm{erf}\left(\frac{\ln(t) + \ln\left(\frac{\sum_{i=1}^{n}\frac{t_i - t_{i-1}}{e^{\mu_i}}}{t}\right)}{\sqrt{2\sigma^2}}\right) \tag{2.18}$$

where the scale parameter μ_i for the ith interval is

$$\mu_i = \ln\left(\mathrm{MTTF_{WO}}(T_i)\right) - \frac{\sigma^2}{2} \tag{2.19}$$

The reliability of a core with N devices using Lognormal distribution for fault density is given by [22]

$$R(t) = e^{N\int_{-\infty}^{\infty} f(x)\ln\left(\frac{1}{2} - \frac{1}{2}\mathrm{erf}\left(\frac{\ln(t)-\mu}{\sqrt{2x^2}}\right)\right)dx} \tag{2.20}$$

where $f(x)$ is the probability density function of μ. This equation is too complex to integrate in the reliability computation for systems. Furthermore, recent studies reveal that for large number of devices per component, Weibull distribution provides more accurate modeling than Lognormal distribution and therefore has been adopted for most of the existing works on reliability optimization.

2.2.2.3 MTTF of a Core

The mean time to failure for a core c_i is therefore

$$\mathrm{MTTF}_i = \int_0^\infty R(t)dt \tag{2.21}$$

2.2.3 System-Level Reliability Modeling

System-level reliability modeling is to combine reliability of different cores to determine the MTTF of multiprocessor system. Some of the most widely used MTTF computation techniques are highlighted here.

Max–Min Approach One of the most widely adopted approaches for MTTF analysis is the Max–Min approach [5, 11, 21], where MTTF of a system is approximated to the minimum MTTFs of the different cores, i.e.,

$$\mathrm{MTTF_{sys}} = \min_i \mathrm{MTTF}_i \tag{2.22}$$

Algorithm 1 Iterative reliability computation

Require: Application and architecture graphs
Ensure: MTTF of the multiprocessor systems
 1: Initialize $MTTF_{sys} = 0$
 2: Map and schedule the application on the system
 3: **while** performance requirement is satisfied **do**
 4:     **for all** core c_i of the system **do**
 5:         Determine $MTTF_i$
 6:     **end for**
 7:     $MTTF_{sys} = MTTF_{sys} + \min\{MTTF_i\}$
 8:     Task migration and determine new schedule
 9: **end while**

Additive Refinement Approach In the iterative approach, MTTF is determined iteratively, considering one failure at a time [9]. After every failure, the tasks on the faulty core are remapped to the other active cores. This changes the operating temperature and hence the shape parameter of the fault distribution functions. The new MTTF are computed for the cores and minimum MTTF of all cores is added to system MTTF. This process is repeated until the performance of an application drops below an acceptable limit. Algorithm 1 provides the pseudo-code of the approach.

Multi-Convolution Integral Approach The reliability of a multiprocessor system with l failures is given by [10]

$$R_{N_c-l}^{sys}(t) = \int_0^t dt_1 \int_{t_1}^t dt_2 \cdots \int_{t_{l-1}}^t R_{N_c-l}^{sys}(t, \mathbf{t}_l) dt_l \qquad (2.23)$$

where N_c is the number of cores of the multiprocessor system, l is the number of failures, $\mathbf{t}_l = (t_1, t_2, \cdots, t_l)$, where t_i is the occurrence of the ith failure and

$$R_{N_c-l}^{sys}(t, \mathbf{t}_l) = R_{N_c-l}^{sys}(t|\mathbf{t}_l) \cdot g_{N_c}^{sys}(t_1) \cdot g_{N_c-1}^{sys}(t_2|t_1) \cdots g_{N_c-l+1}^{sys}(t_l|t_1, t_2, \cdots t_{l-1}) \qquad (2.24)$$

where $R_{N_c-l}^{sys}(t|\mathbf{t}_l)$ is the probability that a core survives at time t given the system experiences l failures and $g_{N_c-r+1}^{sys}(t_r|t_1, t_2, \cdots t_{r-1})$ is the probability that a system containing $N_c - r + 1$ working cores experiences the rth failure at time t_r with the past $r - 1$ failures occurring at $t_1, t_2, \cdots t_{r-1}$. The MTTF of the system is

$$\mathrm{MTTF_{sys}} = \int_0^\infty \sum_{l=N_c^{min}}^{N_c} R_{N_c-l}^{sys}(t) dt \qquad (2.25)$$

where N_c^{min} is the minimum number of cores required to satisfy the performance of a system. For a gracefully degrading system, $N_c^{min} = 1$.

Monte-Carlo Simulation Approach The MTTF of a system can be derived using Monte-Carlo simulation using survival lattice to describe system structure. The time to failure of a component assuming Weibull distribution is given by [13, 16, 22]

$$\text{TTF} = \frac{t}{\sum_{i=1}^{n} \frac{t_i - t_{i-1}}{\eta_i}} e^{\frac{\mu - \sqrt{\mu^2 + 2\sigma \ln\left(-\frac{\ln(1-u)}{N_c}\right)}}{\sigma^2}} \qquad (2.26)$$

where u is a uniform random number in $[0, 1]$ representing the expected lifetime during system-level simulation. System-level MTTF is determined using the survival lattice. The system reliability can be calculated as the percentage of trials for which system survives longer than TTF.

References

1. M. Alam, A critical examination of the mechanics of dynamic NBTI for PMOSFETs, in *IEEE International Electron Devices Meeting (IEDM)* (2003), pp. 14.4.1–14.4.4
2. M. Alam, S. Mahapatra, A comprehensive model of {PMOS} {NBTI} degradation. Microelectron. Reliab. **45**(1), 71–81 (2005)
3. J.S.S.T. Association, Failure Mechanisms and Models for Semiconductor Devices (JEDEC Publication JEP122-B, 2003)
4. J.R. Black, Electromigration failure modes in aluminum metallization for semiconductor devices. Proc. IEEE **57**(9), 1587–1594 (1969)
5. T. Chantem, X. Hu, R. Dick, Temperature-aware scheduling and assignment for hard real-time applications on MPSoCs. IEEE Trans. Very Large Scale Integr. Syst. (TVLSI) **19**(10), 1884–1897 (2011)
6. K. Chopra, C. Zhuo, D. Blaauw, D. Sylvester, A statistical approach for full-chip gate-oxide reliability analysis, in *Proceedings of the International Conference on Computer Aided Design (ICCAD)* (IEEE Press, 2008), pp. 698–705
7. S. Downing, D. Socie, Simple rainflow counting algorithms. Int. J. Fatigue **4**(1), 31–40 (1982)
8. A. Ghamarian, M. Geilen, S. Stuijk, T. Basten, A. Moonen, M. Bekooij, B. Theelen, M. Mousavi, Throughput analysis of synchronous data flow graphs, in *Proceedings of the International Conference on Application of Concurrency to System Design (ACSD)*, vol. 6 (IEEE Computer Society, 2006), pp. 25–36
9. Z. Gu, C. Zhu, L. Shang, R. Dick, Application-specific MPSoC reliability optimization. IEEE Trans. Very Large Scale Integr. Syst. (TVLSI) **16**(5), 603–608 (2008)
10. L. Huang, Q. Xu, Lifetime reliability for load-sharing redundant systems with arbitrary failure distributions. IEEE Trans. Reliab. **59**(2), 319–330 (2010)
11. L. Huang, F. Yuan, Q. Xu, On task allocation and scheduling for lifetime extension of platform-based MPSoC designs. IEEE Trans. Parallel Distrib. Syst. (TPDS) **22**(12), 2088–2099 (2011)
12. E. Lee, D. Messerschmitt, Synchronous data flow. Proc. IEEE **75**(9), 1235–1245 (1987)
13. A. Miele, Lifetime reliability modeling and estimation in multi-core systems, in *34th IEEE VLSI Test Symposium, VTS 2016, Las Vegas, NV, April 25–27* (2016), p. 1
14. V. Radhakrishnan, On the bilinearity of the coffin-manson low-cycle fatigue relationship. Int. J. Fatigue **14**(5), 305–311 (1992)
15. T. Shimokawa, S. Tanaka, A statistical consideration of miner's rule. Int. J. Fatigue **2**(4), 165–170 (1980)

16. J. Srinivasan, S.V. Adve, P. Bose, J.A. Rivers, The case for lifetime reliability-aware micro-processors, in *Proceedings of the Annual International Symposium on Computer Architecture (ISCA)* (IEEE Computer Society, 2004), pp. 276–287
17. S. Sriram, S. Bhattacharyya, *Embedded Multiprocessors; Scheduling and Synchronization* (Marcel Dekker, New York, 2000)
18. S. Stuijk, M. Geilen, T. Basten, Exploring trade-offs in buffer requirements and throughput constraints for synchronous dataflow graphs, in *Proceeding of the Annual Design Automation Conference (DAC)* (ACM, 2006), pp. 899–904
19. E. Takeda, N. Suzuki, An empirical model for device degradation due to hot-carrier injection. IEEE Electron Device Lett. **4**(4), 111–113 (1983)
20. S. Tam, P.-K. Ko, C. Hu, Lucky-electron model of channel hot-electron injection in MOS-FET'S. IEEE Trans. Electron Devices **31**(9), 1116–1125 (1984)
21. I. Ukhov, M. Bao, P. Eles, Z. Peng, Steady-state dynamic temperature analysis and reliability optimization for embedded multiprocessor systems, in *Proceeding of the Annual Design Automation Conference (DAC)* (ACM, 2012), pp. 197–204
22. Y. Xiang, T. Chantem, R.P. Dick, X.S. Hu, L. Shang, System-level reliability modeling for MPSoCs, in *Proceedings of the Conference on Hardware/Software Codesign and System Synthesis (CODES+ISSS)* (ACM, 2010), pp. 297–306
23. T. Yamamoto, K. Uwasawa, T. Mogami, Bias temperature instability in scaled p+ polysilicon gate p-MOSFET's. IEEE Trans. Electron Devices **46**(5), 921–926 (1999)

Chapter 3
Literature Survey on System-Level Optimizations Techniques

3.1 Design-Time Based Reliability and Energy Optimization

As discussed in Chap. 1, design-time methodologies address three aspects—reliability and energy-aware platform-based design, reliability and energy-aware hardware–software co-design, and energy-aware mapping for proactive fault-tolerance. Existing studies on these three aspects are discussed next.

3.1.1 Task Mapping Approaches for Platform-Based Design

Research activities on task mapping for platform-based design can be broadly classified into three categories: approaches focusing solely on energy optimization, those on temperature optimization, and those on lifetime reliability optimization. These are discussed next.

Energy Optimization A technique is proposed in [40] to map tasks/IP cores on a generic NoC-based MPSoC. The objective is to minimize energy consumption under performance constraints. Energy efficient task mapping technique is presented in [63] for off-line DVFS enabled embedded systems. A design space exploration framework is proposed in [1] to obtain energy minimum task mappings. This approach called EWARDS explores performance and power capabilities of modern homogeneous MPSoCs at design-time using Model-Driven Engineering (MDE) techniques. A multi-objective heuristic is proposed in [45]. This approach uses multi-objective evolutionary algorithms with low-level backfilling heuristics to map workflows into resources maximizing quality of service, while reducing the computation energy. A task mapping technique is proposed for hybrid main memory in [8]. The proposed approach uses integer linear programming (ILP) to obtain the optimal task allocation for DRAM and nonvolatile memories. A topology aware

© Springer International Publishing AG 2018
A.K. Das et al., *Reliable and Energy Efficient Streaming Multiprocessor Systems*,
Embedded Systems, https://doi.org/10.1007/978-3-319-69374-3_3

mapping approach is proposed in [12] to minimize energy consumption on the communication links of supercomputers. A quantum inspired mapping approach is proposed in [13] to reduce energy consumption of heterogeneous computing systems. An energy-aware scheduling algorithm is proposed in [2] for massively parallel systems. A leakage power minimization technique is proposed in [3], which determines the best task mapping to reduce the static power consumption. For an overview of other approaches for energy optimization, readers are referred to [65].

Thermal Optimization Since temperature has a significant impact on device wear-out, there are quite some studies on off-line task allocation techniques for temperature minimization. A thermal-aware task mapping technique is proposed in [44] based on *HotSpot* tool. A mixed integer linear programming (MILP)-based task mapping and scheduling is proposed in [9] that solves steady-state and spatial temperature dependency from resistive capacitive (RC) thermal equivalent model. A fast event-driven approach is proposed in [17] to estimate temperature of a multiprocessor system using pre-built look-up tables and predefined leakage calibration parameters. All these techniques determine steady-state temperature only. A temperature-aware technique is proposed in [59, 73] to distribute idle time in order to control the power consumption and hence the temperature. Both transient and steady-state phases are modeled in these approaches. However, spatial dependency is not considered. Sixteen heuristics are compared and contrasted in [64]. These heuristics are compared in terms of performance, energy, and temperature for task mapping on a multiprocessor system. The thermal model used in this approach is hugely simplified to reduce design space exploration time.

Wear-Out Optimization Since different wear-out mechanisms are influenced by temperature differently, there are also studies that optimize lifetime reliability, directly considering these wear-out mechanisms through intelligent task mapping. A task mapping technique is proposed in [60] to map dependent tasks on a multiprocessor system. The approach is based on a simplified reliability model. A dynamic fault-tolerance approach is proposed in [11] for mapping tasks on a multiprocessor system. Another reliability-aware task mapping technique is proposed in [36]. This approach uses the reliability models proposed in [58]. A reliability estimation technique is proposed in [35] for application specific multiprocessor systems. The model considers multiple failures by incorporating changes in fault density with core failures. A slack allocation technique is proposed in [57] to improve lifetime reliability of NoC-based multiprocessor systems. This technique exploits critical quantity slack arising from execution and storage resources to increase the MTTF. A simulated annealing based technique is proposed in [42] to maximize the lifetime reliability of a multiprocessor system. Steady-state temperature values are determined using the *HotSpot* tool for all combinations of active tasks on different processors. These temperature data are stored in a lookup and used during the optimization step. Ant-colony based optimization technique is proposed in [38] to determine task mapping that maximizes lifetime defined as the time to the first failure. This technique has shown that the lifetime of a multiprocessor system using temperature-aware optimization technique can be significantly lower than when lifetime is explicitly optimized. A simulated annealing based energy-reliability joint optimization technique is proposed in [41] based on the temperature model of [42].

The technique in [68, 69] uses eigenvalue decomposition based approach to determine steady-state dynamic temperature profile using time-varying and periodic power profiles. The approach is shown to be the most accurate among all the existing techniques. Based on this temperature model, a simple heuristic is proposed to maximize the lifetime of a multiprocessor system considering thermal cycling-related wear-out effects. Finally, there are also techniques to optimize the lifetime of multiprocessor systems considering transient and intermittent faults. A time-series analysis technique is proposed for intermittent faults in [61]. Based on this, an efficient task mapping technique is proposed for multiprocessor systems. A resource management technique is proposed in [14] to minimize processor wear-outs, simultaneously providing tolerance for transient and intermittent faults. A genetic algorithm based lifetime optimization technique is proposed in [27]. This approach determines voltage and frequency of cores of a multiprocessor system to maximize its lifetime and minimize soft-error susceptibility. Markov-decision based multiprocessor steady-state availability is derived in [20] considering intermittent faults. Table 3.1 summarizes these related works and highlights contribution of this book.

Table 3.1 Related works on reliability and energy-aware platform-based design

Related works	Temperature model	Optimization objective	Reliability modeling	Application model	Architecture model
Gu et al. [35]	Steady-state & spatial	Lifetime reliability	*Additive refinement*	Independent DAGs	Static heterogeneous
Meyer et al. [57], Huang et al. [42], Hartman et al. [38]	Steady-state & spatial	Lifetime reliability	*Max-Min*	Independent DAGs	Static homogeneous
Huang et al. [41]	Steady-state & spatial	Lifetime reliability & energy	*Max-Min*	Independent DAGs	Static homogeneous
Ukhov et al. [68]	Transient, steady-state, temporal & spatial	Lifetime reliability & energy	*Max-Min*	Independent DAGs	Static homogeneous
Chou et al. [14], Das et al. [20, 27]	Steady-state & spatial	Lifetime reliability, transient & intermittent faults	*Max-Min*	Independent DAGs	Static homogeneous
Proposed [21, 23, 28]	Transient, steady-state, temporal & spatial	Lifetime reliability & energy	*Additive refinement*	Independent & concurrent SDFGs	Static homogeneous

3.1.2 Existing Approaches on Hardware–Software Co-design

A hardware–software co-design approach is proposed in [47] for control applications with real-time and reliability constraints. A hardware–software co-synthesis of fault-tolerant systems is proposed in [29]. The proposed approach uses task duplication with comparison and re-execution as the fault-tolerant mechanism. Online soft-error mitigation technique is proposed for preemptable tasks in [53]. The approach is also based on hardware–software partitioning. A design methodology is proposed in [71] to handle conditional execution in multi-rate embedded systems and selectively duplicates critical tasks to correct transient errors. A technique is proposed in [6] to determine cost, performance, and reliability trade-offs for multiprocessor system considering permanent faults. A design space exploration of multimedia multiprocessor systems is proposed in [66]. The proposed approach explores the trade-off between different metrics such as performance, energy, and cost while incorporating soft-error tolerance in the optimization process.

A system-level reliability analysis technique is proposed in [46] considering process re-execution in software and selective hardening of hardware for fault-tolerance. Based on this, a design optimization heuristic is proposed to select the fault-tolerant architecture and the task mapping such that the overall cost is minimized. A system-level synthesis flow is proposed in [5] for the design of reliable embedded systems. The methodology explores different hardening strategies under a given user level reliability specification. The strategy with least resource utilization is selected and target applications are mapped on the resulting platform to optimize reliability. All these techniques determine multiprocessor platform to maximize the fault-tolerance capability considering transient and permanent faults.

There are very few research studies on the hardware–software co-design methodology for proactive fault-tolerance considering temperature-related wear-out. A system-level design methodology is proposed in [34] for the automatic synthesis of reliable embedded systems. This methodology addresses the following: selection of resources with different reliability, area and latency parameters; and mapping of a data flow application on the platform to simultaneously optimize reliability, area and latency using multi-objective evolutionary algorithm. A co-design methodology is proposed in [74] to determine the minimum resources required to improve system lifetime measured as MTTF. Table 3.2 summarizes the related works and the contributions of this book.

3.1.3 Existing Approaches on Reactive Fault-Tolerance

The off-line reactive fault-tolerant techniques address task allocation problem on a multiprocessor system considering permanent faults. A multi-objective optimization approach is proposed in [49] to jointly optimize cost, time, and dependability. Application task graph is extended with some nodes of the graph replicated to

Table 3.2 Related works on reliability and energy-aware hardware–software co-design

Related works	Optimization objective	Reliability modeling	Fault-tolerance	Multiprocessor platform	Application model
Dave et al. [29], Xie et al. [71], Bolchini et al. [5]	Area	–	Reactive	Static heterogeneous	Independent DAGs
Bolchini et al. [6], Stralen et al. [66], Izosimov et al. [46]	Reliability	–	Reactive	Static homogeneous	Independent DAGs
Glaß et al. [34]	Lifetime reliability	*Convolution integral*	Proactive	Static homogeneous	Independent DAGs
Zhu et al. [74]	Area & lifetime reliability	*Additive refinement*	Proactive	Static heterogeneous	Independent DAGs
Proposed [22]	Lifetime reliability	*Max-Min*	Proactive & reactive	Homogeneous & reconfigurable	Independent & concurrent SDFGs

allow fault-tolerance. The dependability of the replicated task graph is evaluated considering resource crash as a fault model. This approach performs design space exploration using replicated graph in a genetic algorithm framework to maximize the dependability, while satisfying design cost and execution time constraints. A fixed order Band-and-Band reconfiguration technique is studied in [72]. Cores of the target architecture are partitioned into two bands. When one or more cores fail, tasks on these core(s) are migrated to other functional core(s) determined by the band in which these tasks belong. Core partitioning strategy is fixed at design-time and is independent of application throughput requirements. A re-execution slot based reconfiguration mechanism is studied in [43]. Normal and re-execution slots of a task are scheduled at design-time using evolutionary algorithm to minimize certain parameters like throughput degradation. At run-time, tasks on a faulty core migrate to their re-execution slot on a different core. However, a limitation of this technique is that schedule length can become unbounded for high fault-tolerant systems. Task remapping technique based on off-line computation and virtual mapping is proposed in [52]. Here, task mapping is performed in two steps—determining the highest throughput mapping followed by the generation of a virtual mapping to minimize the cost of task migration to achieve this highest throughput mapping. An ILP based approach is presented in [70]. Energy optimization is performed under execution time constraint which incorporates fault-tolerance overhead using check-pointing based recovery model. Table 3.3 summarizes the existing reactive fault-tolerant techniques and the proposed approach.

Table 3.3 Related works on design-time reactive fault-tolerant techniques

Related works	Fault-tolerant mapping & scheduling	Energy optimization	Migration overhead	Application model	Throughput degradation
Jhumka et al. [49]	Task mapping only	×	×	DAGs	×
Yang et al. [72]	Task mapping only	×	✓	DAGs	×
Huang et al. [43]	Mapping & scheduling	×	×	DAGs	×
Lee et al. [52]	Task mapping only	×	×	DAGs	✓
Wei et al. [70]	Task mapping only	✓	×	DAGs	×
Proposed [18, 19, 24, 25]	Mapping & scheduling	✓	✓	SDFGs	✓

3.2 Run-Time Based Reliability and Energy Optimization

Run-time optimization techniques minimize energy and maximize reliability based on dynamisms of the workload executed on the system at a given point in time. A graceful performance degradation approach is proposed in [67] for MPSoC providing fault-tolerance at run-time. Real-time task allocation is proposed in [55] for video decoding application on NoC. Another real-time energy-aware task mapping approach is proposed in [54] for communication tasks on CPU-GPU hybrid clusters. Similar to this approach, the technique in [30] performs run-time mapping of general tasks for energy optimization.

A dynamic reliability management technique is proposed in [50], where workload characteristics and thermal information are used to project degradation caused by various failure mechanisms. The relationship between temperature, voltage, and frequency of operation is formulated in [37]. Based on this, an online heuristic is proposed to determine voltage and frequency of cores to minimize temperature. In [32], a distributed dynamic thermal management technique is proposed to avoid thermal hot spots that accelerate thermal aging and transient faults. This technique consists of multiple agents, each managing a cluster of the many-core architecture. A run-time task mapping technique is proposed in [39] to minimize wear-out by utilizing wear-out sensors. The technique computes task mapping at regular intervals and also when a component fails. Scheduling decisions are left to the operating system. An online approach is proposed in [10] to minimize wear-out considering different aging mechanisms. Power consumption of a core is used to estimate its temperature and a simple heuristic is proposed to dynamically manage peak temperature and thermal cycling. Both these techniques are based on simulation of a real multiprocessor system.

A proactive thermal management policy is proposed in [16] to balance temperature on the die. This approach uses auto-regressive moving average modeling to forecast future temperature. A proactive dynamic thermal management is proposed in [4] based on predict-and-act philosophy. In the framework, operating system scheduler predicts temperature of individual cores; if this temperature crosses a predefined value, the operating system decides to migrate one or more threads of a

Table 3.4 Related works on run-time reliability optimization techniques

Related works	DRM approach	Workload variation	Thermal cycling	Temperature/reliability measurements	Validation platform
Karl et al. [50], Hanumaiah et al. [37], Faruque et al. [32]	Heuristic	Intra variation	×	Thermal model	Multi-core
Hartman et al. [39]	Heuristic	Intra variation	×	Wear-out sensors	Simulation
Chantem et al. [10]	Heuristic	Intra variation	✓	Thermal model	Simulation
Coskun et al. [16], Ayoub et al. [4], Cochran et al. [15]	Predict-and-act	Intra variation	×	Thermal sensors	Multi-core
Sironi et al. [62]	Observed-decide-act	Intra variation	×	Thermal sensors	Multi-core
Bolchini et al. [7]	Observe-decide-act	Intra variation	×	*HotSpot*	Simulation
Mercati et al. [56]	Observe-decide-act	Intra variation	✓	Wear-out sensors	Multi-core
Lee et al. [51]	Machine learning	Intra variation	×	Sensors	Multi-core
Jayaseelan et al. [48]	Machine learning	Intra variation	×	*HotSpot*	Simulation
Ge et al. [33]	Machine learning	Intra variation	×	Thermal models & sensors	Multi-core
Ebi et al. [31]	Machine learning	Intra variation	×	Thermal gun	FPGA
Proposed [26]	Machine learning	Inter & intra variation	✓	Thermal sensors	Multi-core

given workload to the coolest core in the system. A thermal management technique is proposed in [15], which predicts workload phase change and selects appropriate voltage and frequency of cores to minimize peak temperature. A discrete-time thermal model is proposed in [62] for dynamic thermal management. This technique monitors temperature sensor and decides on the length of idle time needed to reduce thermal emergencies. An adaptive approach is proposed in [7] to minimize electromigration-related wear-out by monitoring aging at run-time and controlling it through task mapping. A control theoretic approach is proposed in [56] to maximize lifetime of homogeneous multi-core systems. The approach is based on long-term controller, which samples data from aging sensors to compute wear-out degradation. Based on this, a short-term controller adjusts voltage and frequency of the tasks to minimize temperature while satisfying performance requirement.

A machine learning technique is proposed in [51] to dynamically manage peak temperature for MPEG2 video decoding. The technique is application specific and can be applied only to video decoding applications such as MPEG4 or H.264. A neural network-based adaptive thermal management policy is proposed in [48]. The technique relies on temperature prediction using the *HotSpot* tool. A reinforcement learning algorithm is proposed in [33] to manage performance-thermal trade-offs by sampling temperature data from on-board thermal sensors. A distributed learning agent is proposed in [31] to optimize peak temperature within a given power budget. The technique is implemented on FPGA with temperature measurement using an external thermal gun. Table 3.4 summarizes these related works.

References

1. M. Ammar, M. Baklouti, M. Pelcat, K. Desnos, M. Abid, Off-line DVFS integration in mde-based design space exploration framework for mp2soc systems, in *IEEE 25th International Conference on Enabling Technologies: Infrastructure for Collaborative Enterprises (WETICE)* (IEEE, 2016), pp. 160–165
2. M. Ammar, M. Baklouti, M. Pelcat, K. Desnos, M. Abid, On exploiting energy-aware scheduling algorithms for MDE-based design space exploration of mp2soc, in *2016 24th Euromicro International Conference on Parallel, Distributed, and Network-Based Processing (PDP)* (IEEE, 2016), pp. 643–650
3. G. Ananthanarayanan, S.R. Sarangi, M. Balakrishnan, Leakage power aware task assignment algorithms for multicore platforms, in *2016 IEEE Computer Society Annual Symposium on VLSI (ISVLSI)* (IEEE, 2016), pp. 607–612
4. R.Z. Ayoub, T.S. Rosing, Predict and act: dynamic thermal management for multi-core processors, in *Proceedings of the ACM/IEEE International Symposium on Low Power Electronics and Design (ISLPED)* (ACM, 2009), pp. 99–104
5. C. Bolchini, A. Miele, Reliability-driven system-level synthesis of embedded systems, in *Proceedings of the IEEE International Symposium on Defect and Fault Tolerance in VLSI and Nanotechnology Systems (DFT)*, (Oct 2010), pp. 35–43
6. C. Bolchini, A. Miele, F. Salice, D. Sciuto, L. Pomante, Reliable system co-design: the FIR case study, in *Proceedings of the IEEE International Symposium on Defect and Fault Tolerance in VLSI and Nanotechnology Systems (DFT)* (Oct 2004), pp. 433–441
7. C. Bolchini, M. Carminati, A. Miele, A. Das, A. Kumar, B. Veeravalli, Run-time mapping for reliable many-cores based on energy/performance trade-offs, in *Proceedings of the IEEE International Symposium on Defect and Fault Tolerance in VLSI and Nanotechnology Systems (DFT)* (2013), pp. 58–64
8. X. Cai, L. Ju, X. Li, Z. Zhang, Z. Jia, Energy efficient task allocation for hybrid main memory architecture. J. Syst. Archit. **71**, 12–22 (2016)
9. T. Chantem, X. Hu, R. Dick, Temperature-aware scheduling and assignment for hard real-time applications on MPSoCs. IEEE Trans. Very Large Scale Integr. Syst. (TVLSI) **19**(10), 1884–1897 (2011)
10. T. Chantem, Y. Xiang, X.S. Hu, R.P. Dick, Enhancing multicore reliability through wear compensation in online assignment and scheduling. in *Proceedings of the Conference on Design, Automation and Test in Europe (DATE)* (European Design and Automation Association, 2013), pp. 1373–1378
11. N. Chatterjee, S. Paul, S. Chattopadhyay, Fault-tolerant dynamic task mapping and scheduling for network-on-chip-based multicore platform. ACM Trans. Embedded Comput. Syst. (TECS) **16**(4), 108 (2017)

12. J. Chen, Y. Tang, Y. Dong, J. Xue, Z. Wang, W. Zhou, Reducing static energy in supercomputer interconnection networks using topology-aware partitioning. IEEE Trans. Comput. **65**(8), 2588–2602 (2016)
13. S. Chen, Z. Li, B. Yang, G. Rudolph, Quantum-inspired hyper-heuristics for energy-aware scheduling on heterogeneous computing systems. IEEE Trans. Parallel Distrib. Syst. **27**(6), 1796–1810 (2016)
14. C.-L. Chou, R. Marculescu, FARM: fault-aware resource management in noc-based multiprocessor platforms, in *Proceedings of the Conference on Design, Automation and Test in Europe (DATE)* (European Design and Automation Association, 2011), pp. 1–6
15. R. Cochran, S. Reda, Consistent runtime thermal prediction and control through workload phase detection, in *Proceeding of the Annual Design Automation Conference (DAC)* (ACM, 2010), pp. 62–67
16. A. Coskun, T. Rosing, K. Gross, Utilizing predictors for efficient thermal management in multiprocessor SoCs. IEEE Trans. Comput. Aided Des. Integr. Circuits Syst. (TCAD) **28**(10), 1503–1516 (2009)
17. J. Cui, D. Maskell, A fast high-level event-driven thermal estimator for dynamic thermal aware scheduling. IEEE Trans. Comput. Aided Des. Integr. Circuits Syst. (TCAD) **31**(6), 904–917 (2012)
18. A. Das, A. Kumar, Fault-aware task re-mapping for throughput constrained multimedia applications on NoC-based MPSoCs, in *Proceedings of the International Symposium on Rapid System Prototyping (RSP)* (IEEE, 2012), pp. 149–155
19. A. Das, A. Kumar, B. Veeravalli, Energy-aware communication and remapping of tasks for reliable multimedia multiprocessor systems, in *Proceedings of the International Conference on Parallel and Distributed Systems (ICPADS)* (IEEE Computer Society, 2012), pp. 564–571
20. A. Das, A. Kumar, B. Veeravalli, Communication and migration energy aware design space exploration for multicore systems with intermittent faults, in *Proceedings of the Conference on Design, Automation and Test in Europe (DATE)* (European Design and Automation Association, 2013), pp. 1631–1636
21. A. Das, A. Kumar, B. Veeravalli, Reliability-driven task mapping for lifetime extension of networks-on-chip based multiprocessor systems, in *Proceedings of the Conference on Design, Automation and Test in Europe (DATE)* (European Design and Automation Association, 2013), pp. 689–694
22. A. Das, A. Kumar, B. Veeravalli, Aging-aware hardware-software task partitioning for reliable reconfigurable multiprocessor systems, in *Proceedings of the International Conference on Compilers, Architecturesand Synthesis for Embedded Systems (CASES)* (IEEE Press, 2013), pp. 1:1–1:10
23. A. Das, A. Kumar, B. Veeravalli, Temperature aware energy-reliability trade-offs for mapping of throughput-constrained applications on multimedia MPSoCs, in *Proceedings of the Conference on Design, Automation and Test in Europe (DATE)* (European Design and Automation Association, 2014)
24. A. Das, A. Kumar, B. Veeravalli, Communication and migration energy aware task mapping for reliable multiprocessor systems. Elsevier Futur. Gener. Comput. Syst. **30**, 216–228 (2014)
25. A. Das, A. Kumar, B. Veeravalli, Energy-aware task mapping and scheduling for reliable embedded computing systems. ACM Trans. Embed. Comput. Syst. (TECS) **13**(2s), 72:1–72:27 (2014)
26. A. Das, R.A. Shafik, G.V. Merrett, B.M. Al-Hashimi, A. Kumar, B. Veeravalli, Reinforcement learning-based inter- and intra-application thermal optimization for lifetime improvement of multicore systems, in *Proceeding of the Annual Design Automation Conference (DAC)* (2014)
27. A. Das, A. Kumar, B. Veeravalli, C. Bolchini, A. Miele, Combined DVFS and mapping exploration for lifetime and soft-error susceptibility improvement in MPSoCs, in *Proceedings of the Conference on Design, Automation and Test in Europe (DATE)* (European Design and Automation Association, 2014)

28. A. Das, A. Kumar, B. Veeravalli, Reliability and energy-aware mapping and scheduling of multimedia applications on multiprocessor systems. IEEE Trans. Parallel Distrib. Syst. **27**(3), 869–884 (2016)
29. B. Dave, N. Jha, COFTA: hardware-software co-synthesis of heterogeneous distributed embedded systems for low overhead fault tolerance. IEEE Trans. Comput. **48**(4), 417–441 (1999)
30. B. Donyanavard, T. Mück, S. Sarma, N. Dutt, Sparta: runtime task allocation for energy efficient heterogeneous many-cores, in *Proceedings of the Eleventh IEEE/ACM/IFIP International Conference on Hardware/Software Codesign and System Synthesis* (ACM, 2016), p. 27
31. T. Ebi, D. Kramer, W. Karl, J. Henkel, Economic learning for thermal-aware power budgeting in many-core architectures, in *Proceedings of the Conference on Hardware/Software Codesign and System Synthesis (CODES+ISSS)* (ACM, 2011), pp. 189–196
32. M.A. Faruque, J. Jahn, J. Henkel, Runtime thermal management using software agents for multi- and many-core architectures. IEEE Des. Test Comput. **27**(6), 58–68 (2010)
33. Y. Ge, Q. Qiu, Dynamic thermal management for multimedia applications using machine learning, in *Proceeding of the Annual Design Automation Conference (DAC)* (ACM, 2011), pp. 95–100
34. M. Glaß, M. Lukasiewycz, T. Streichert, C. Haubelt, J. Teich, Reliability-aware System Synthesis, in *Proceedings of the Conference on Design, Automation and Test in Europe (DATE)*, (EDA Consortium, 2007), pp. 409–414
35. Z. Gu, C. Zhu, L. Shang, R. Dick, Application-specific MPSoC reliability optimization. IEEE Trans. Very Large Scale Integr. Syst. (TVLSI) **16**(5), 603–608 (2008)
36. M.-H. Haghbayan, A. Miele, A.M. Rahmani, P. Liljeberg, H. Tenhunen, A lifetime-aware runtime mapping approach for many-core systems in the dark silicon era, in *Design, Automation & Test in Europe Conference & Exhibition (DATE), 2016* (IEEE, 2016), pp. 854–857
37. V. Hanumaiah, S. Vrudhula, Temperature-aware DVFS for hard real-time applications on multicore processors. IEEE Trans. Comput. **61**(10), 1484–1494 (2012)
38. A.S. Hartman, D.E. Thomas, B.H. Meyer, A case for lifetime-aware task mapping in embedded chip multiprocessors, in *Proceedings of the Conference on Hardware/Software Codesign and System Synthesis (CODES+ISSS)* (ACM, 2010), pp. 145–154
39. A.S. Hartman, D.E. Thomas, Lifetime improvement through runtime wear-based task mapping, in *Proceedings of the conference on hardware/software codesign and system synthesis (CODES+ISSS)* (ACM, 2012), pp. 13–22
40. J. Hu, R. Marculescu, Energy-aware mapping for tile-based NoC architectures under performance constraints, in *Proceedings of the 2003 Asia and South Pacific Design Automation Conference* (ACM, 2003), pp. 233–239
41. L. Huang, Q. Xu, Energy-efficient task allocation and scheduling for multi-mode MPSoCs under lifetime reliability constraint, in *Proceedings of the Conference on Design, Automation and Test in Europe (DATE)* (European Design and Automation Association, 2010), pp. 1584–1589
42. L. Huang, F. Yuan, Q. Xu, On task allocation and scheduling for lifetime extension of platform-based MPSoC designs. IEEE Trans. Parallel Distrib. Syst. (TPDS) **22**(12), pp. 2088–2099 (2011)
43. J. Huang, J.O. Blech, A. Raabe, C. Buckl, A. Knoll, Analysis and optimization of fault-tolerant task scheduling on multiprocessor embedded systems, in *Proceedings of the Conference on Hardware/Software Codesign and System Synthesis (CODES+ISSS)* (ACM, 2011), pp. 247–256
44. W.-L. Hung, Y. Xie, N. Vijaykrishnan, M. Kandemir, M.J. Irwin, Thermal-aware task allocation and scheduling for embedded systems, in *Proceedings of the Conference on Design, Automation and Test in Europe (DATE)* (IEEE Computer Society, 2005), pp. 898–899
45. S. Iturriaga, B. Dorronsoro, S. Nesmachnow, Multiobjective evolutionary algorithms for energy and service level scheduling in a federation of distributed datacenters. Int. Trans. Oper. Res. **24**(1–2), 199–228 (2017)

46. V. Izosimov, I. Polian, P. Pop, P. Eles, Z. Peng, Analysis and optimization of fault-tolerant embedded systems with hardened processors, in *Proceedings of the Conference on Design, Automation and Test in Europe (DATE)* (European Design and Automation Association, 2009), pp. 682–687

47. B. Janßen, M. Naserddin, M. Hübner, A hardware/software co-design approach for control applications with static real-time reallocation, in *2016 IEEE International Parallel and Distributed Processing Symposium Workshops* (IEEE, 2016) pp. 241–246

48. R. Jayaseelan, T. Mitra, Dynamic thermal management via architectural adaptation, in *Proceeding of the Annual Design Automation Conference (DAC)* (ACM, 2009), pp. 484–489

49. A. Jhumka, S. Klaus, S.A. Huss, A dependability-driven system-level design approach for embedded systems, in *Proceedings of the Conference on Design, Automation and Test in Europe (DATE)* (IEEE Computer Society, 2005), pp. 372–377

50. E. Karl, D. Blaauw, D. Sylvester, T. Mudge, Reliability modeling and management in dynamic microprocessor-based systems, in *Proceeding of the Annual Design Automation Conference (DAC)* (ACM, 2006), pp. 1057–1060

51. W. Lee, K. Patel, M. Pedram, GOP-level dynamic thermal management in MPEG-2 decoding. IEEE Trans. Very Large Scale Integr. Syst. (TVLSI) **16**(6), 662–672 (2008)

52. C. Lee, H. Kim, H.-W. Park, S. Kim, H. Oh, S. Ha, A task remapping technique for reliable multi-core embedded systems, in *Proceedings of the Conference on Hardware/Software Codesign and System Synthesis (CODES+ISSS)* (ACM, 2010), pp. 307–316

53. A. Martínez-Álvarez, F. Restrepo-Calle, S. Cuenca-Asensi, L.M. Reyneri, A. Lindoso, L. Entrena, A hardware-software approach for on-line soft error mitigation in interrupt-driven applications. IEEE Trans. Dependable Secure Comput. **13**(4), 502–508 (2016)

54. X. Mei, X. Chu, H. Liu, Y.-W. Leung, Z. Li, Energy efficient real-time task scheduling on cpu-gpu hybrid clusters, in *INFOCOM. IEEE* (2017)

55. H.R. Mendis, N.C. Audsley, L.S. Indrusiak, Dynamic and static task allocation for hard real-time video stream decoding on NoCs. Leibniz Trans. Embedded Syst., 1–25 (2017). http://eprints.whiterose.ac.uk/117785/

56. P. Mercati, A. Bartolini, F. Paterna, T.S. Rosing, L. Benini, Workload and user experience-aware dynamic reliability management in multicore processors, in *Proceeding of the Annual Design Automation Conference (DAC)* (ACM, 2013), pp. 2:1–2:6

57. B.H. Meyer, A.S. Hartman, D.E. Thomas, Cost-effective slack allocation for lifetime improvement in NoC-based MPSoCs, in *Proceedings of the Conference on Design, Automation and Test in Europe (DATE)* (European Design and Automation Association, 2010), pp. 1596–1601

58. A. Miele, Lifetime reliability modeling and estimation in multi-core systems, in *34th IEEE VLSI Test Symposium, VTS 2016, Las Vegas, NV, April 25–27* (2016), p. 1

59. D. Rai, H. Yang, I. Bacivarov, J.-J. Chen, L. Thiele, Worst-case temperature analysis for real-time systems, in *Proceedings of the Conference on Design, Automation and Test in Europe (DATE)* (2011), pp. 1–6

60. V. Rathore, V. Chaturvedi, T. Srikanthan, Performance constraint-aware task mapping to optimize lifetime reliability of manycore systems, in *2016 International Great Lakes Symposium on VLSI* (IEEE, 2016), pp. 377–380

61. S.S. Sahoo, A. Kumar, B. Veeravalli, Design and evaluation of reliability-oriented task re-mapping in mpsocs using time-series analysis of intermittent faults, in *2016 Design, Automation & Test in Europe Conference & Exhibition (DATE)* (IEEE, 2016), pp. 798–803

62. F. Sironi, M. Maggio, R. Cattaneo, G. Del Nero, D. Sciuto, M. Santambrogio, ThermOS: system support for dynamic thermal management of chip multi-processors, in *Proceedings of the International Conference on Parallel Architectures and Compilation Techniques (PACT)* (2013), pp. 41–50

63. M.T. Schmitz, B.M. Al-Hashimi, P. Eles, Energy-efficient mapping and scheduling for DVS enabled distributed embedded systems, in *Proceedings of Design, Automation and Test in Europe Conference and Exhibition, 2002* (IEEE, 2002), pp. 514–521

64. H.F. Sheikh, I. Ahmad, Sixteen heuristics for joint optimization of performance, energy, and temperature in allocating tasks to multi-cores. ACM Trans. Parallel Comput. (TOPC) **3**(2), 9 (2016)
65. A.K. Singh, M. Shafique, A. Kumar, J. Henkel, Mapping on multi/many-core systems: survey of current and emerging trends, in *Proceedings of the 50th Annual Design Automation Conference* (ACM, 2013), p. 1
66. P.V. Stralen, A. Pimentel, A SAFE approach towards early design space exploration of fault-tolerant multimedia MPSoCs, in *Proceedings of the Conference on Hardware/Software Codesign and System Synthesis (CODES+ISSS)* (ACM, 2012), pp. 393–402
67. S. Tzilis, I. Sourdis, V.Vasilikos, D. Rodopoulos, D. Soudris, Runtime management of adaptive mpsocs for graceful degradation, in *Proceedings of the International Conference on Compilers, Architectures and Synthesis for Embedded Systems* (ACM, 2016), p. 5
68. I. Ukhov, M. Bao, P. Eles, Z. Peng, Steady-state dynamic temperature analysis and reliability optimization for embedded multiprocessor systems, in *Proceeding of the Annual Design Automation Conference (DAC)* (ACM, 2012), pp. 197–204
69. I. Ukhov, P. Eles, Z. Peng, Probabilistic analysis of electronic systems via adaptive hierarchical interpolation. IEEE Trans. Comput. Aided Des. Integr. Circuits Syst. **36**(11), 1883–1896 (2017). https://doi.org/10.1109/TCAD.2017.2705117
70. T. Wei, X. Chen, S. Hu, Reliability-driven energy-efficient task scheduling for multiprocessor real-time systems. IEEE Trans. Comput. Aided Des. Integr. Circuits Syst. (TCAD) **30**(10), 1569–1573 (2011)
71. Y. Xie, L. Li, M. Kandemir, N. Vijaykrishnan, M. Irwin, Reliability-aware co-synthesis for embedded systems. J. VLSI Signal Process. Syst. Signal Image Video Technol. **49**(1), 87–99 (2007)
72. C. Yang, A. Orailoglu, Predictable execution adaptivity through embedding dynamic reconfigurability into static MPSoC schedules, in *Proceedings of the Conference on Hardware/Software Codesign and System Synthesis (CODES+ISSS)* (ACM, 2007), pp. 15–20
73. J. Zhou, J. Yan, J. Chen, T. Wei, Peak temperature minimization via task allocation and splitting for heterogeneous mpsoc real-time systems. J. Signal Process. Syst. **84**(1), 111–121 (2016)
74. C. Zhu, Z. P. Gu, R.P. Dick, L. Shang, Reliable multiprocessor system-on-chip synthesis, in *Proceedings of the Conference on Hardware/Software Codesign and System Synthesis (CODES+ISSS)* (ACM, 2007), pp. 239–244

Chapter 4
Reliability and Energy-Aware Platform-Based Multiprocessor Design

4.1 Introduction

As discussed in Chap. 3, a significant research is conducted recently to investigate platform-based design approaches in order to mitigate wear-out and minimize energy consumption. These studies, however, suffer from two limitations:

Accuracy Most prior works on thermal and reliability management ignore the transient phase of temperature or the spatial dependency. Ignoring the transient phase simplifies thermal analysis, but is accurate if only the execution time of applications or tasks is comparable to thermal time constant of the package (typically few hundreds of seconds). Let us consider the example of playback of a video encoded at 24 frames per second. The decoding time per frame is approximately 42 ms, which is much lower than the thermal time constraint ($\approx 100\,\mathrm{s}$). Assuming steady-state temperature for each frame decoding interval is clearly not accurate.

Ignoring the spatial dependency leads to simplification of the RC equivalent model, but results in underestimation of temperature which overestimates the mean time to failure (MTTF). Additionally, some prior studies on lifetime reliability approximate MTTF as the time to first fault. This is true for systems that are not provisioned to tolerate faults. In this chapter, multiprocessor systems are considered with support for task migration. Such a system continues to operate in presence of faults, albeit an acceptable performance degradation. For such systems, estimating the MTTF as time to the first failure leaves a significant scope of improvement, both in terms of lifetime and energy consumption.

Scope Existing lifetime optimization techniques are based on sequential execution of applications modeled as directed acyclic graphs (DAGs). Synchronous data flow graphs (SDFGs) allow more suitable modeling for streaming multimedia and other data flow applications that require support for multi-input tasks, multi-rate tasks, and pipelined execution. Techniques for DAGs cannot be applied directly on SDFGs due to cyclic actor dependencies and overlapping of multiple iterations (pipelined)

© Springer International Publishing AG 2018

A.K. Das et al., *Reliable and Energy Efficient Streaming Multiprocessor Systems*,
Embedded Systems, https://doi.org/10.1007/978-3-319-69374-3_4

in the schedule. Furthermore, all existing techniques determine lifetime optimum mapping for a single application. Modern platforms are designed to support multiple applications enabled concurrently (use-case). As shown in this chapter, lifetime-aware distribution of cores among concurrent applications leads to a significant improvement in MTTF.

To address these limitations, a temperature model is first proposed that is based on off-line thermal characterization of a multiprocessor system using the *HotSpot* tool [12]. The model incorporates the following:

1. *temporal dependency* i.e., relationship between temperature of a core, the operating voltage-frequency and time; and
2. *spatial dependency* i.e., influence of neighboring core's temperature on the temperature of a core.

A gradient-based fast heuristic is proposed incorporating the aforementioned temperature model, to jointly optimize energy and lifetime reliability of a multiprocessor system with applications modeled as SDFG. This approach leverages on the SDF^3 tool [13]. Following are the key contributions of this chapter:

- a simplified temperature-model considering temporal phase and spatial dependencies;
- computing MTTF considering task remapping;
- a gradient-based fast heuristic to jointly optimize lifetime reliability and energy;
- reliability optimization considering SDFGs; and
- MTTF maximization considering single and multi-application use-cases.

The remainder of this chapter is organized as follows. The optimization problem is formulated in Sect. 4.2 followed by description of our proposed temperature model in Sect. 4.3. Temperature computation from a given SDFG schedule is demonstrated in Sect. 4.4. The proposed design methodology is discussed next in Sect. 4.5. Experimental results are presented in Sect. 4.6 and conclusions in Sect. 4.7.

4.2 Problem Formulation

4.2.1 Application Model

An application SDFG is mathematically represented as $\mathcal{G}_{app} = (\mathbb{A}, \mathscr{C})$ consisting of a finite set $\mathbb{A}$ of actors and a finite set $\mathscr{C}$ of channels. Every actor $\boldsymbol{a}_i \in \mathbb{A}$ is a tuple (t_i, μ_i), where t_i is the execution time of $\boldsymbol{a}_i$ and μ_i is its state space (program and data memory). The number of actors in an SDFG is denoted by N_a where $N_a = |\mathbb{A}|$. Performance of an SDFG is specified in terms of throughput constraint $\mathbb{T}_c$.

Fig. 4.1 An example
multiprocessor architecture

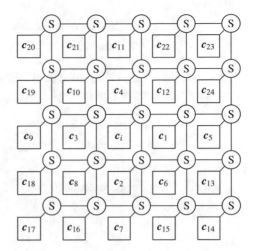

4.2.2 Architecture Model

An example multiprocessor architecture is shown in Fig. 4.1 with cores (identified by labeled boxes) interconnected through switches (identified by labeled circles) in a mesh topology (interpretation of these labels is provided in Sect. 4.3). This model is a representative of a single chip multi-/many-core system.

For problem formulation, the above architecture is represented as a graph $\mathscr{G}_{arc} = (\mathbb{C}, \mathbb{E})$, where $\mathbb{C}$ is the set of nodes representing cores of the architecture and $\mathbb{E}$ is the set of edges representing communication channels among cores. The number of cores in the architecture is denoted by N_c, i.e., $N_c = |\mathbb{C}|$. Each core $c_j \in \mathbb{C}$ supports N_f voltage-frequency pairs denoted by $\{(V_k, \omega_k) \ \forall k \in [0, N_f - 1]\}$.

4.2.3 Mapping Representation

The objective of the optimization problem is to maximize lifetime reliability (measured as MTTF) and minimize the energy consumption by solving the following:

- *actor distribution:* assign actors of an application SDFG on cores of a multiprocessor system;
- *operating point:* assign voltage and frequency of cores for executing these actors.

For problem formulation, two variables $x_{i,j}$ (representing actor distribution) and $y_{i,k}$ (representing operating point) are defined as follows.

$$x_{i,j} = \begin{cases} 1 & \text{if actor } a_i \text{ is executed on core } c_j \\ 0 & \text{otherwise} \end{cases}$$

$$y_{i,k} = \begin{cases} 1 & \text{if actor } \boldsymbol{a}_i \text{ is executed at operating point } (V_k, \omega_k) \\ 0 & \text{otherwise} \end{cases}$$

These variables are constrained such that an actor is always mapped to only one core at a single operating point, i.e.,

$$\sum_{j=0}^{N_c-1} x_{i,j} = 1 \text{ and } \sum_{k=0}^{N_f-1} y_{i,k} = 1 \quad \forall \boldsymbol{a}_i \in \mathbb{A} \tag{4.1}$$

Actor distribution and operating point of SDFG are represented as two matrices:

$$\mathcal{M}_d = \begin{pmatrix} x_{0,0} & x_{0,1} & \cdots & x_{0,N_c-1} \\ x_{1,0} & x_{1,1} & \cdots & x_{1,N_c-1} \\ \vdots & \vdots & \ddots & \vdots \\ x_{N_a-1,0} & x_{N_a-1,1} & \cdots & x_{N_a-1,N_c-1} \end{pmatrix} \tag{4.2}$$

$$\mathcal{M}_o = \begin{pmatrix} y_{0,0} & y_{0,1} & \cdots & y_{0,N_f-1} \\ y_{1,0} & y_{1,1} & \cdots & y_{1,N_f-1} \\ \vdots & \vdots & \ddots & \vdots \\ y_{N_a-1,0} & y_{N_a-1,1} & \cdots & y_{N_a-1,N_f-1} \end{pmatrix} \tag{4.3}$$

Core assignment for actor $\boldsymbol{a}_i$ is denoted by ϕ_i and can be calculated as $\phi_i = \mathbf{X}_i \times \mathbb{N}_{N_c}$ where $\mathbf{X}_i = \begin{pmatrix} x_{i,0} & x_{i,1} & \cdots & x_{i,N_c-1} \end{pmatrix}$ and $\mathbb{N}_{N_c}$ is the matrix of integers from 0 to N_c, i.e., $\mathbb{N}_{N_c} = \begin{pmatrix} 0 & 1 & \cdots & N_c - 1 \end{pmatrix}^T$. The operating point of actor $\boldsymbol{a}_i$ is denoted by θ_i and is given by $\theta_i = \mathbf{Y}_i \times \mathbb{N}_{N_f}$ where $\mathbf{Y}_i = \begin{pmatrix} y_{i,0} & y_{i,1} & \cdots & y_{i,N_f-1} \end{pmatrix}$.

4.2.4 MTTF Computation

To demonstrate the MTTF computation using the proposed iterative approach (Chap. 2), an example is provided with three cores. The initial schedule $\mathscr{S}_0$ uses all three cores and stresses core 2 more than the other two cores. Reliability curves for the three cores are shown in Fig. 4.2. Core 2 has the least lifetime and it fails at time τ_0. As indicated in Chap. 2, prior works on lifetime reliability define MTTF to be the time to the first failure, hence the MTTF for these works is τ_0. At time $t = \tau_0$, a new schedule $\mathscr{S}_1$ (using core 0 and 1) is applied. Changes in reliability profiles of core 0 and core 1 are due to different wear-outs, is due to difference in temperatures from the newly applied schedule. This new schedule stresses core 0 more than core 1 and therefore core 0 fails at time $t = \tau_1$. At this time, all actors are remapped to

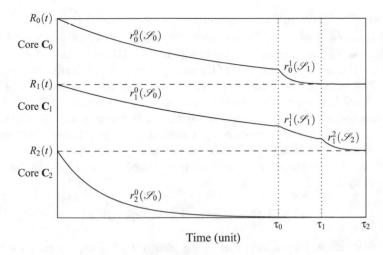

Fig. 4.2 MTTF computation with different temperature profile

core 1 and are ordered honoring actor dependency. This schedule is identified in the figure as $\mathscr{S}_2$ and results in reliability profile of r_1^2. With this new reliability profile, core 1 fails at time $t = \tau_2$. The lifetime (MTTF) of the system is therefore τ_2.

An important metric for multiprocessor systems supporting task remapping is the *processor years*, defined as the aggregate utilization of different cores of the system over its lifetime. For the above example, this is calculated as follows. For the interval 0 to τ_0, all three cores are active; for the interval τ_0 to τ_1 two cores are active; and for the interval τ_1 to τ_2 only one core is active. Assuming the time in this figure is all in years, the *processor years* of the above system is $3 \cdot \tau_0 + 2 \cdot (\tau_1 - \tau_0) + 1 \cdot (\tau_2 - \tau_1)$.

4.2.5 Energy Computation

Actor-level voltage and frequency scaling is assumed, i.e., every actor of an SDFG is associated with a voltage-frequency value that is set on the core executing this actor. Total energy consumed on the multiprocessor system consists of the following:

- *computation energy:* dynamic and leakage energy consumed on the cores due to actor execution; and
- *communication energy:* dynamic and leakage energy consumed on the network-on-chip (NoC) due to data communication among connected actors.[1]

---

[1] Spatial division multiplexing-based NoC is assumed in this work and therefore, the leakage power consumed on the NoC is negligible [10].

Dynamic Energy of SDFG Dynamic energy of an SDFG is given by $E^{\text{dyn}} = E_{\text{tr}}^{\text{dyn}} + N_{\text{iter}} \cdot E_{\text{ss}}^{\text{dyn}}$, where $E_{\text{tr}}^{\text{dyn}}$ is the actor dynamic energy in the transient phase of the schedule, $E_{\text{ss}}^{\text{dyn}}$ is the actor dynamic energy per iteration of the steady-state phase, and N_{iter} is the number of iterations of the steady-state phase. Usually, the number of steady state iterations (i.e., N_{iter}) is a large number (e.g., periodic decoding of video frames during a video playback) and hence for practical purposes, dynamic energy of the steady-state phase dominates over that in the transient phase. Throughout the rest of this chapter, computation (or communication) energy implies computation (or communication) energy of the steady-state phase per iteration.

Dynamic energy consumed by an actor a_i executed on core c_j at operating point k is given by

$$e^{\text{dyn}}(i, j, k) = C_{\text{eff}} \cdot \beta \cdot V_k^2 \cdot \omega_k \cdot t_{ijk} \cdot RV[a_i] \tag{4.4}$$

where β is the activity factor, C_{eff} is the effective load capacitance, t_{ijk} is the execution time of actor a_i on core c_j at operating point k (i.e.,operating voltage V_k and operating frequency ω_k), and $RV[a_i]$ is the number of firings of actor a_i per steady-state iteration of the SDFG. The total dynamic energy per steady-state iteration of the SDFG is

$$E_{\text{core}}^{\text{dyn}} = \sum_{\forall a_i \in \mathbb{A}} e^{\text{dyn}}(i, \phi_i, \theta_i) \tag{4.5}$$

Leakage Energy of SDFG The leakage energy of core c_j, consumed during the execution of actor a_i at operating point k, is given by the following equation [11].

$$e^{\text{leak}}(i, j, k) = N_{\text{gates}} V_k I_0 \left[AT^2 e^{\frac{\alpha V_k + \beta}{T}} + Be^{\gamma V_k + \delta} \right] \cdot t_{ijk} \cdot RV[a_i] \tag{4.6}$$

where N_{gates} is the number of gates of the core, I_0 is the average leakage current and $A, B, \alpha, \beta, \gamma, \delta$ are technology dependent constants (refer to [11]), and T is the average temperature of the actor during the steady-state iteration. The total leakage energy is

$$E_{\text{core}}^{\text{leak}} = \sum_{\forall a_i \in \mathbb{A}} e^{\text{leak}}(i, \phi_i, \theta_i) \tag{4.7}$$

Dynamic Energy on NoC In [7], bit energy (E_{bit}) is defined as the energy consumed in transmitting one bit of data through routers and links of a NoC.

$$E_{\text{bit}} = E_{S_{\text{bit}}} + E_{L_{\text{bit}}} \tag{4.8}$$

where $E_{S_{\text{bit}}}$ and $E_{L_{\text{bit}}}$ are energy consumed in the switch and the link, respectively. Energy per bit consumed in transferring data between cores c_p and c_q, situated $n_{\text{hops}}(p, q)$ away is given by Eq. (4.9) according to [7].

$$E_{\text{bit}}(p, q) = \begin{cases} \left(n_{\text{hops}}(p, q) + 1 \right) \cdot E_{S_{\text{bit}}} + n_{\text{hops}}(p, q) \cdot E_{L_{\text{bit}}} & \text{if } p \neq q \\ 0 & \text{otherwise} \end{cases} \quad (4.9)$$

Dynamic energy consumed on the NoC is therefore given by Eq. (4.10), where ϕ_i and $\phi_{i'}$ are cores where actors a_i and $a_{i'}$ are mapped, respectively.

$$E_{\text{noc}}^{\text{dyn}} = \sum_{\forall a_i, a_{i'} \in A} d_{ij} \cdot E_{\text{bit}}(\phi_i, \phi_{i'}) \quad (4.10)$$

The total energy is

$$E^{\text{tot}} = E_{\text{core}}^{\text{dyn}} + E_{\text{core}}^{\text{leak}} + E_{\text{noc}}^{\text{dyn}} \quad (4.11)$$

4.2.6 Reliability-Energy Joint Metric

Lifetime reliability and energy are combined into a single metric **lifetime quotient** (*lq*), which is defined as the ratio of MTTF to the total energy, i.e.,

$$lq = \frac{MTTF}{E^{\text{tot}}} \quad (4.12)$$

The optimization objective is to maximize *lq*.

4.3 Proposed Temperature Model

Temperature of a multiprocessor system can be calculated by solving a RC equivalent thermal model. Temperature of a single core is related to its power dissipation according to the following equation [12].

$$C\frac{dT(t)}{dt} + G\left(T(t) - T_{\text{amb}}\right) = P(t) \quad (4.13)$$

where C is the thermal capacitance, G is the thermal conductance, t is the time, T_{amb} is the ambient temperature, $T(t)$ is the instantaneous temperature, and $P(t)$ is the instantaneous power that is composed of dynamic and leakage components. The dynamic power depends on voltage and frequency of operation and the leakage power depends on the temperature (refer to Sect. 4.2.5). Solution to the above equation consists of transient and steady-state phases. In the transient phase, temperature increases with time up to a point beyond which the steady-state phase settles in and temperature saturates to its steady-state value.

For a multiprocessor system with interconnected cores (refer to Fig. 4.1), temperature of any core, say core c_i depends on

A.1   Time of execution of an actor on c_i.
A.2   Voltage and frequency of c_i.
A.3   Temperature of the cores surrounding c_i.

A.1 and A.2 represent temporal phase and A.3 represents the spatial dependency. For such a system, temperature, power, thermal capacitance, and thermal conductance in Eq. (4.13) are all vectors. The transient and steady-state values can be obtained by solving the above equation analytically as

$$\mathbf{T}(t) = e^{\kappa t}\mathbf{T}(0) + \kappa^{-1}\left(e^{\kappa t} - \mathbf{I}\right)\mathbf{C}^{-1}\mathbf{P}(t) \tag{4.14}$$

where $\kappa = -\mathbf{C}^{-1}\mathbf{G}$, $\mathbf{T}(0)$ is the initial temperature and $\mathbf{I}$ is the identity matrix. Direct solution techniques, such as LU decomposition and sparse solver, are usually slow and result in an exponential design space exploration time. Although iterative techniques of [14] simplify the solution, execution time is still exponential when applied to multi-application use-cases.

One approach to simplify the analytical formulation is to ignore spatial dependency by considering temperature for cores individually, such as that proposed in [1]. To show the importance of temperature underestimation obtained by ignoring the spatial dependency (component A.3), an experiment is conducted using the *HotSpot* tool to measure the steady-state temperature. The multiprocessor architecture for the *HotSpot* tool is shown in Fig. 4.1 with specifications of cores reported in Table 4.1. Temperature is recorded by setting the power dissipation of core c_i to 0, and setting a constant power at operating point OPP1G (i.e., 1.35 V, 1 GHz) for all one-hop and two-hop neighboring cores (cores c_1–c_4 are the one-hop neighbors and cores c_5–c_{12} are the two-hop neighbors of core c_i in Fig. 4.1). This arrangement simulates a scenario where core c_i is idle with neighboring cores active at highest voltage and frequency. Temperature results are reported in Fig. 4.3 for some combinations of the neighboring core's activity. Labels $c_1 - c_n$ in the figure indicate cores $c_1, c_2, \cdots, c_n$ are active simultaneously. There are two bars shown on the plot. The left bar for each label corresponds to the temperature of core c_i obtained with all other cores idle. The right bar corresponds to the temperature of c_i with cores $c_1, c_2, \cdots, c_n$ operating at the highest operating point and core c_i as idle. As can be seen from this figure, with only the east and the south neighbors active (i.e.,label

Table 4.1 ARM processor specification

Power mode	Frequency	Voltage (V)	Current (mA)	Power (mW)
OPP50	300 MHz	0.93	151.62	141.01
OPP100	600 MHz	1.10	328.79	361.67
OPP130	800 MHz	1.26	490.61	618.17
OPP1G	1 GHz	1.35	649.64	877.01

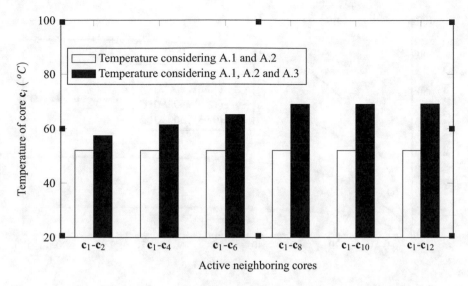

Fig. 4.3 Temperature underestimation ignoring the spatial dependency

$c_1 - c_2$), temperature considering spatial dependency is 5°C higher than temperature obtained by ignoring the spatial dependency (i.e., A.3). This difference increases as more neighbors become active. Finally, with all one-hop and two-hop neighbors active, the temperature difference is 18°C. This temperature underestimation leads to MTTF misprediction.

We propose a regression technique to solve Eq. (4.14). The proposed temperature model is based on:

- temperature characterization to incorporate temporal dependency (capturing both transient and steady-state behaviors); and
- temperature characterization to incorporate spatial temperature dependency.

Solution to the differential equation is represented as

$$T_i(t) = f(V_i, \omega_i, t) + g(\{V_j, \omega_j \mid \forall c_j \in \aleph(c_i)\}) \qquad (4.15)$$

where (V_i, ω_i) are voltage and frequency of core c_i, t is the time, and $\aleph(c_i)$ are cores in the neighborhood of c_i. The function f and g represent temporal and spatial dependencies, respectively, and are derived in two steps.

- **Determine f**: The function f can be determined using one of two alternative approaches—(1) by solving Eq. (4.13) directly for a processing core; or (2) by simulation using the *HotSpot* tool for power consumption corresponding to different operating points of the processing core to capture the transient and the steady-state behaviors, as shown in Fig. 4.4.

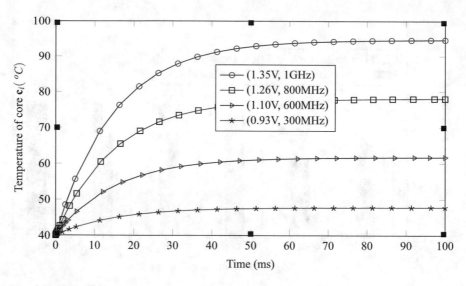

Fig. 4.4 Characterization to determine temporal dependency

- **Characterize g**: The temperature data for characterizing the function g are obtained as follows. The core c_i is set to idle and the operating points for the neighboring cores are altered. Performing exhaustive temperature simulations for different voltage-frequency combinations of all active neighbors is time consuming, but only required once during the characterization step. A first order of approximation involves considering the voltage-frequency of only the immediate neighbors i.e., the east, west, north, and south neighbors of a core referred to as (V_e, ω_e), (V_w, ω_w), (V_n, ω_n), and (V_s, ω_s), respectively with all other neighbors set to operate at the highest operating point. This is shown in Fig. 4.5. The figure plots temperature of core c_i as its voltage V_i is increased from 0.93 to 1.35 V for few of these neighboring voltage combinations.

These thermal data are fed to Matlab regression toolbox to derive the temperature model. The final expression for temperature (Eq. (4.15)) can be easily integrated inside a design space exploration framework. However, the proposed model incorporates pessimism in three forms—separating temporal and spatial dependency; characterizing spatial dependency with steady-state temperature of the nearest neighbors; and characterizing spatial dependency with non-nearest neighbors set to operate at the highest voltage and frequency. These pessimism lead to a temperature overestimation by as much as up to 6°C. However, as discussed in Sect. 4.5.2, this temperature overestimation simplifies reliability optimization for multi-application use-cases commonly executed on multiprocessor systems.

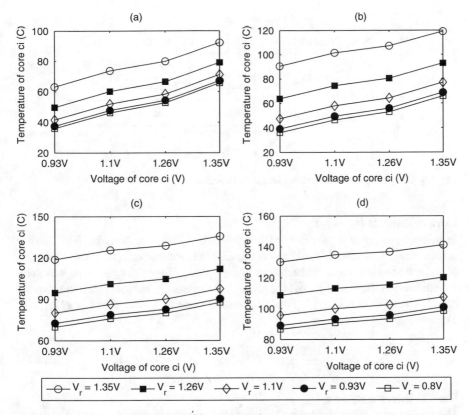

Fig. 4.5 Characterization to determine spatial dependency. (**a**) $V_n = V_w = V_s = 0, V_e = V_r$. (**b**) $V_n = V_w = 0, V_s = V_e = V_r$. (**c**) $V_n = 0, V_w = V_s = V_e = V_r$. (**d**) $V_n = V_w = V_s = V_e = V_r$

4.4   Computing Temperature from a Schedule

Figure 4.6 shows an example SDFG with four actors allocated on a platform with three cores. Schedule corresponding to a particular allocation is also shown in the same figure. Temperature computation is demonstrated for core 0 using this schedule. Temperature for other cores can be determined in a similar manner. The time duration $0 - t_6$ is divided into seven intervals shown by dotted lines at instances where actors start or end *firing*.

Core 0: Interval $(0 \rightarrow t_0)$

In this interval, core 0 executes actor $\boldsymbol{a}$ at operating point (V_A, ω_A). Temperature at time t considering temporal effect is $f(V_A, \omega_A, t)$. Temperature considering the spatial effect is due to idle voltages of cores 1 and 2 and is given by $g(V_{\text{idle}}, V_{\text{idle}})$. Average temperature in this interval is

$$T_0(0, t_0) = \frac{1}{t_0} \int_0^{t_0} f(V_A, \omega_A, t)dt + g(V_{\text{idle}}, V_{\text{idle}}) \qquad (4.16)$$

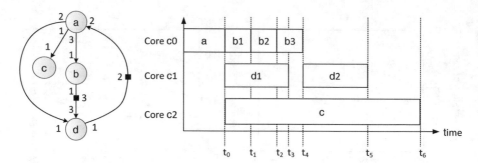

Fig. 4.6 Computing temperature from an SDFG schedule

Core 0: Interval $(t_0 \rightarrow t_1)$
In this interval, core 0 executes the first instance of actor b. Note in SDFG, when actor a fires, it produces 3 tokens on the channel from actor a to actor b and one of these tokens is consumed each time actor b fires. There are three firings of actor b (indicated in the figure by b_1, b_2, and b_3). Temperature at time t due to temporal effect of actor b is $f(V_B, \omega_B, t)$ and temperature due to spatial effect is $g(V_D, V_C)$. The average temperature is

$$T_0(t_0, t_1) = \frac{1}{t_1 - t_0} \int_0^{t_1 - t_0} f(V_B, \omega_B, t)dt + g(V_D, V_C) \qquad (4.17)$$

Core 0: Interval $(t_1 \rightarrow t_2)$
The temperature computation in this interval is similar to that in the interval $(t_0 \rightarrow t_1)$ and is given by

$$T_0(t_1, t_2) = \frac{1}{t_2 - t_1} \int_0^{t_2 - t_1} f(V_B, \omega_B, t)dt + g(V_D, V_C) \qquad (4.18)$$

Core 0: Interval $(t_2 \rightarrow t_3)$
During the execution of actor b_3, there is a change in temperature profile due to completion of actor d_1 and the interval before actor d_2 is executed. Hence, execution time of actor b_3 is split into two intervals $(t_2 \rightarrow t_3)$ and $(t_3 \rightarrow t_4)$. Temperature computation in the interval $(t_2 \rightarrow t_3)$ is similar to that in the interval $(t_0 \rightarrow t_1)$

$$T_0(t_2, t_3) = \frac{1}{t_3 - t_2} \int_0^{t_3 - t_2} f(V_B, \omega_B, t)dt + g(V_D, V_C) \qquad (4.19)$$

Core 0: Interval $(t_3 \rightarrow t_4)$
The average temperature in this interval is given by

$$T_0(t_3, t_4) = \frac{1}{t_4 - t_3} \int_0^{t_4 - t_3} f(V_B, \omega_B, t)dt + g(V_{\text{idle}}, V_C) \qquad (4.20)$$

Core 0: Interval $(t_4 \rightarrow t_5)$
In this interval, the temporal effect is due to idle temperature of the core and is denoted by T_0^{idle}. The average temperature is given by

$$T_0(t_4, t_5) = T_0^{\text{idle}} + g(V_D, V_C) \tag{4.21}$$

Core 0: Interval $(t_5 \rightarrow t_6)$
The temperature in this interval is given by

$$T_0(t_5, t_6) = T_0^{\text{idle}} + g(V_{\text{idle}}, V_C) \tag{4.22}$$

Reliability of Core 0
Combining these equations, aging of core 0 is

$$r_0 = \frac{1}{t_6} \sum_{i=0}^{6} \frac{t_i - t_{i-1}}{\alpha \left(T_0(t_i, t_{i-1}) \right)} \tag{4.23}$$

where $t_{-1} = 0$ and α is the fault density.

4.5   Design Methodology

The proposed design methodology consists of two phases—analysis at design-time (application and use-case optimizations) and execution at run-time. The design-time methodology is highlighted in Fig. 4.7. Run-time manager is described in details in Chap. 7 and is shown here for the sake of completeness. There are two databases for the multiprocessor system—the set of applications ($\mathbf{S}_{\text{app}}$) and the set of use-cases ($\mathbf{S}_{\text{use}}$). The proposed approach is to determine actor distribution and operating points (refer to Sect. 4.2) for every application using $n = N_c^{\text{min}}$ to N_c cores of the system. Thus, $|\mathbf{S}_{\text{app}}| \cdot (N_c - N_c^{\text{min}} + 1)$ optimization problems are solved at design-time inside the block labeled *REOpt*. The solution consists of actor distribution and operating point matrices stored in the *MapDB* database and the three-dimensional (3D) vector—throughput, reliability (MTTF), and core count stored in the *ThRiCoDB* database.

Algorithm 2 provides pseudo-code of this design methodology. For every application $\mathbf{A}_i$ of the set $\mathbf{S}_{\text{app}}$, the corresponding SDFG representation and throughput constraint are fetched from the database. This application is executed on the multiprocessor system with n cores identified as $\mathscr{G}_{\text{arc}}^n$, where n is varied from N_c^{min} to N_c. The reliability-energy joint optimization is first performed on the application (line 5) to obtain the actor distribution matrix $\mathscr{M}_d$ and the operating point matrix $\mathscr{M}_o$. These are stored in the *MapDB* (line 7). The actor distribution is used in the

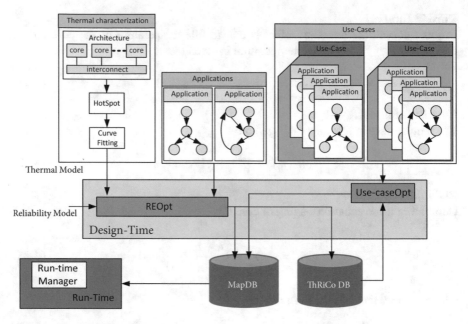

Fig. 4.7 Proposed design flow

Algorithm 2 Generate reliability and energy aware mappings

Require: Application set $\mathbf{S}_{app}$ and multiprocessor system $\mathcal{G}_{arc}$
Ensure: *MapDB* and *ThRiCoDB*

1: **for all** Application $\mathbf{A}_i \in \mathbf{S}_{app}$ **do**
2:      $[\mathcal{G}_{app} \;\; \mathbb{T}_c] = GetSDFG(\mathbf{A}_i)$ *//Get the corresponding SDFG and the throughput constraint.*
3:      Determine N_c^{min}, the minimum number of cores required for satisfying the throughput requirement.
4:      **for** $n = N_c^{min}$ to N_c **do**
5:          $(\mathcal{M}_d \; \mathcal{M}_o) = REOpt(\mathcal{G}_{app}, \mathcal{G}_{arc}^n, \mathbb{T}_c)$ *//Perform individual application optimization*
6:          $[\mathcal{S} \;\; \mathbb{T}] = MSDF^3(\mathcal{M}_d, \mathcal{M}_o, \mathcal{G}_{app}, \mathcal{G}_{arc}^n)$ *//Calculate the schedule and the throughput of the SDFG.*
7:          $MapDB(i, n) = (\mathcal{M}_d \; \mathcal{M}_o \; \mathcal{S})$ *//Store the mapping in the mapping database.*
8:          $M = CalculateMTTF(i, n, N_c^{min}, MapDB)$
9:          $ThRiCoDB(i, n) = (\mathbb{T} \; M)$ *//Store the throughput and MTTF values for the use-case optimization step.*
10:     **end for**
11: **end for**

MSDF^3 tool that leverages on the SDF^3 tool [13][2] to generate the throughput and schedule. The schedule is then used in the *CalculateMTTF()* routine (line 8) to

---

[2]The SDF^3 tool generates one feasible actor distribution and the corresponding throughput. The MSDF^3 tool is modified form of SDF^3 that generates the schedule and throughput from a given actor distribution matrix.

Algorithm 3 *CalculateMTTF*(): Calculate the mean time to failure

---

Require: Application id i, the core index n, the minimum number of cores for throughput satisfaction and mapping database *MapDB*

Ensure: MTTF M

1: Initialize $ttf = 0$ and $ri = n$
2: **while** $ri \geq N_c^{min}$ **do**
3:     $[\mathcal{M}_d \; \mathcal{M}_o \; \mathcal{S}] = MapDB(i, ri)$ //Fetch the values.
4:     Determine reliability profiles from $\mathcal{S}$ as demonstrated in Sect. 4.4
5:     Shift the reliability profiles by ttf
6:     Determine t, the time to failure of the most stressed core
7:     $ttf = ttf + t$ and $ri = ri - 1$
8: **end while**

---

compute the MTTF. Throughput and MTTF values corresponding to the number of cores are stored in the *ThRiCoDB* for the use-case optimization step that addresses core distribution among concurrent applications.

The *CalculateMTTF* routine determines MTTF iteratively as shown as pseudo-code in Algorithm 3. A running index ri is maintained to index to the schedule with one less core. At the start of the iteration, mapping and scheduling are fetched from the *MapDB*. This schedule is used to compute the reliability profile of every core in the system. The reliability profile is shifted to account for the aging already encountered in the cores. The time-to-failure for all cores is determined using Eq. (2.21). The minimum time corresponds to the failure of the most stressed core. This is added to the *ttf* and the running index is decremented.

4.5.1   Reliability Optimization for Individual Application

The objective function lifetime quotient is non-linear; a gradient-based fast heuristic is proposed to solve it. This is shown as pseudo-code in Algorithm 4. The algorithm starts from an initial allocation, computed using the SDF^3 tool (line 2). Subsequently, the algorithm remaps every actor to every core to determine a priority function defined as

$$\mathscr{P} = \begin{cases} \frac{lq_n - lq}{\mathbb{T} - \mathbb{T}_n} & \text{if } \mathbb{T}_n < \mathbb{T} \\ (lq_n - lq) & \text{otherwise} \end{cases} \qquad (4.24)$$

Two conditions are factored in the priority computation: if throughput of the current allocation ($\mathbb{T}_n$) is lower than the original throughput ($\mathbb{T}$), a gradient function is used to calculate its priority i.e., assignments that increase the lifetime quotient with the least throughput degradation are given higher priorities. Conversely, if the current throughput is higher than the original one, high priorities are given to assignments with the largest increase in the lifetime quotient.

Algorithm 4 *REOpt*(): Reliability and energy optimization for an application

---

Require: $\mathscr{G}_{app}, \mathscr{G}_{arc}$ and throughput constraint $\mathbb{T}_c$

Ensure: actor distribution and operating point matrices $(\mathscr{M}_d \; \mathscr{M}_o)$, which maximize lq

  1: Initialize $\mathscr{M}_o = (0 \; 0 \; \cdots \; 1)$ //Initialize the actors to the highest operating point.

  2: $[\mathscr{M}_d \; \mathscr{S} \; \mathbb{T}] = SDF^3(\mathscr{G}_{app}, \mathscr{G}_{arc})$ //Mapping, schedule and throughput using the native SDF^3 tool.

  3: **while** true **do**

  4:     $\mathscr{P}^{best} = 0, \mathscr{M}_d^{best} = \mathscr{M}_d, best\_found = false$ //Initialize the best values.

  5:     $lq = CalculateLQ(\mathscr{M}_d, \mathscr{M}_o, \mathscr{S}, \mathbb{T})$ //Calculate the initial lifetime quotient.

  6:     **for all** $a_i \in \mathbb{A}$ **do**

  7:         **for all** $c_j \in \mathbb{C}$ **do**

  8:             **for all** $k \in [0, N_f - 1)$ **do**

  9:                 $\mathscr{M}_d^{temp} = \mathscr{M}_d$ and $\mathscr{M}_o^{temp} = \mathscr{M}_o$ //A temporary allocation matrix is used.

10:                 Update $\mathscr{M}_d^{temp}, \mathscr{M}_o^{temp}$ using $x_{i,j} = y_{i,k} = 1$ and $x_{i,l} = y_{i,m} = 0, \forall l \neq j$ and $\forall m \neq k$

11:                 $[\mathscr{S}_n \; \mathbb{T}_n] = MSDF^3(\mathscr{M}_d^{temp}, \mathscr{M}_o^{temp}, \mathscr{G}_{app}, \mathscr{G}_{arc})$ //New schedule is computed.

12:                 $lq_n = CalculateLQ(\mathscr{M}_d^{temp}, \mathscr{M}_o^{temp}, \mathscr{S}_n, \mathbb{T})$ //Calculate the new lifetime quotient.

13:                 Compute $\mathscr{P}$ using Eq. (4.24)

14:                 **if** $\mathbb{T}_n > \mathbb{T}_c$ and $\mathscr{P} > \mathscr{P}^{best}$ **then**

15:                     $\mathscr{P}^{best} = \mathscr{P}, \mathscr{M}_d^{best} = \mathscr{M}_d^{temp}, \mathscr{M}_o^{best} = \mathscr{M}_o^{temp}, best\_found = true, \mathbb{T} = \mathbb{T}_n$

16:                 **end if**

17:             **end for**

18:         **end for**

19:     **end for**

20:     **if** *best_found* **then**

21:         $\mathscr{M}_d = \mathscr{M}_d^{best}$ and $\mathscr{M}_o = \mathscr{M}_o^{best}$ //Actor distribution and operating point matrices are updated.

22:     **else**

23:         break

24:     **end if**

25: **end while**

26: Return $(\mathscr{M}_d \; \mathscr{M}_o)$ //Actor distribution and operating point matrices are returned.

The algorithm remaps actor a_i to a core c_j at operating point k (lines 6–8). The actor distribution and the operating point of actor a_i are changed (line 10). These matrices are used by the $MSDF^3$ tool to compute the throughput and schedule corresponding to the allocation $\mathscr{M}_d^{temp}$ (line 11). The *CalculateLQ* function computes the lifetime quotient using Eq. (4.11) to compute the energy and Algorithm 3 to compute the MTTF. The algorithm computes the priority function (line 13). If this priority is greater than the best priority obtained thus far and the throughput constraint is satisfied, the best values are updated (line 15). The algorithm continues as long as an assignment can be found without violating the throughput constraint.

4.5.2   Reliability Optimization for Use-Cases

In this section, the use-case level optimization problem is formulated based on the results obtained in Sect. 4.5.1. It is to be noted that when multiple applications are enabled simultaneously, the temperature due to the execution of one application is dependent not only on the temperature of the cores on which it is executed, but also on the temperature due to other concurrent applications. As a result, wear-out (or the MTTF) due to single application can be significantly different than the actual wear-out (or the MTTF) for use-cases. This limitation is addressed using the pessimism introduced in the thermal model. Specifically, to determine the temperature for different cores during single application mode, all unused cores in the architecture are considered to be operating, and their temperature effect is incorporated in determining the temperature of the actual operating cores. This pessimism simplifies the thermal computation.

As indicated previously, the *ThRiCoDB* contains 3D databases with throughput and MTTF number for every core count of every application. The problem addressed here is to merge these 3D databases for applications enabled simultaneously such that the distribution of the cores among these applications maximizes the system MTTF. For the ease of problem formulation, the following notations are defined:

$$A_1, \cdots, A_n = \text{n applications enabled simultaneously}$$

$$z_i = \text{number of cores for application } A_i$$

$$M_i = \text{MTTF of } A_i \text{ mapped on } z_i \text{ cores} = \textit{ThRiCoDB.getMTTF}(z_i)$$

$$\mathbb{T}_i = \text{Throughput of } A_i \text{ mapped on } z_i \text{ cores} = \textit{ThRiCoDB.getThr}(z_i)$$

4.5.2.1 Problem Formulation

The optimization problem is

$$\text{maximize} \ \min_i \{M_i\}$$

$$\text{subject to} \ \sum_{i=1}^{n} z_i \leq N_c \tag{4.25}$$

$$\forall i, \mathbb{T}_i \geq \text{throughput constraint of } A_i$$

4.5.2.2 Solving Use-Case Level Reliability Minimization

Algorithm 5 provides the pseudo-code to solve Eq. (4.25). A list is defined (*RiList*) to store the applications (their IDs) of the use-case, the number of cores dedicated to it, and the corresponding MTTF value. For every core in the system (line 3), the *RiList* is sorted to determine the application with the least MTTF (lines 4–5). A core is dedicated to this application (line 6); the corresponding MTTF is fetched from the *ThRiCoDB* (line 7), and the *RiList* is updated.

Algorithm 5 Core distribution for use-cases

Require: *ThRiCoDB*
Ensure: Distribution of cores among applications
1: Initialize : $z_i = 0$, $1 \leq i \leq n$
2: Initialize : $RiList.push(A_i, z_i, 0)$, $1 \leq i \leq n$
3: **for** $j = 1$ to N_c **do**
4:     $RiList.sort()$
5:     Let, $A_k = $ Task with least MTTF
6:     $z_k = z_k + 1$
7:     $M_k = ThRiCoDB.getMTTF(A_k, z_k)$
8:     $RiList.update(A_k, z_k, M_k)$
9: **end for**

4.6   Experiments and Discussions

Experiments are conducted with real-life as well as synthetic SDFGs on a multiprocessor system with mesh architecture. Synthetic SDFGs are generated using
the SDF^3 tool [13] with the number of actors ranging between nine and twenty-
five. These applications are a mix of computation and communication dominated
applications. The real-life SDFGs used in the evaluation are *H.263 Encoder, H.263
Decoder, H.264 Encoder, MPEG4 Decoder, JPEG Decoder, MP3 Encoder and
Sample Rate Converter*. Additionally, two non-streaming applications are also
considered to demonstrate the applicability. These are *FFT* and *Romberg Integration*
from [2]. The supported voltage and frequency pairs are reported in Table 4.1, based
on ARM Cortex-A8 core [3]. Although these voltage-frequency pairs are assumed
for simplicity, the proposed algorithm and the thermal model can be used with any
voltage-frequency pairs.

The bit energy (E_{bit}) for modeling the communication energy of an application is calculated using similar methodology as [7] for packet-based NoC with
Batcher-Banyan switch fabric, using 65 nm technology parameters from [15]. The
parameters used for computing the MTTF are similar to those used in [6, 8, 9].
The scale parameter of each core is normalized so that its MTTF under idle (non-
stressed) condition is 10 years. All algorithms are coded in C++, and used with
SDF^3 tool for throughput and schedule construction, and *HotSpot* for thermal
characterization.

4.6.1   Time Complexity

The time complexity of the algorithms is calculated as follows. There are ($N_c -
N_c^{\min} + 1$) loops in the algorithm 2 for each application. In each loop, the algorithm
executes the $REOpt()$, the $MSDF^3()$, and the iterative technique to compute the

MTTF (i.e., Algorithm 3). The complexity of Algorithm 3 is calculated as follows. Assuming lines 3–7 can be computed in unit time, the worst case complexity of this algorithm is

$$C_3 = O\left(N_c\right) \qquad\qquad (4.26)$$

since $N_c^{\min} \leq N_c$. The complexity of *REOpt()* (Algorithm 4) is computed as follows. Let there be η iterations of the outer while loop (lines 3–25). In each iteration, the algorithm maps each actor to each core at each operating point to determine its reliability. The complexity of this algorithm (C_4) is

$$C_4 = O\left(\eta \cdot N_a \cdot N_c \cdot N_f \cdot O(MSDF^3) \cdot C_3\right) \qquad\qquad (4.27)$$

The MSDF^3 engine computes the schedule starting from a given actor distribution. This can be performed in $O\left(N_a \log N_a + N_a \cdot £\right)$ (ref. [5]), where $£$ is the average number of successors of an actor. Therefore,

$$C_4 = O\left(\eta \cdot N_a \cdot N_c \cdot N_f \cdot (N_a \log N_a + N_a \cdot £) \cdot N_c\right) = O\left(N_a^5 \cdot N_f\right) \qquad (4.28)$$

where $N_c, £ \leq N_a$. The overall complexity of the reliability-energy joint optimization for each application is $C_2 = O\left(C_4 + O\left(MSDF^3\right) + C_3\right) = O\left(N_a^5 \cdot N_f\right)$. The execution time of the MSDF^3 tool is reported in Table 4.2.

Finally, the complexity of Algorithm 5 is calculated as follows. For every iteration of the outermost loop (number of cores), sorting of MTTF is performed once followed by the memory lookup. If the memory lookup time is assumed to be constant and there are n applications enabled simultaneously on N_c cores, every loop is executed in $O(n \log n)$. The overall complexity of Algorithm 5 is therefore $O(N_c \times n \log n)$. On the multiprocessor platform considered, this algorithm takes between 80 and 100 μs for two to six simultaneous applications on an architecture with nine homogeneous cores.

Table 4.2 Execution time (s) of the MSDF^3 tool with varying actors and cores

Actors	Multiprocessor platform			
	6 cores	9 cores	12 cores	16 cores
8	3.1	7.6	7.6	7.6
16	6.8	10.1	26.8	101.1
24	217.4	241.7	323.0	409.8
32	899.4	1021.4	2211.0	2789.9

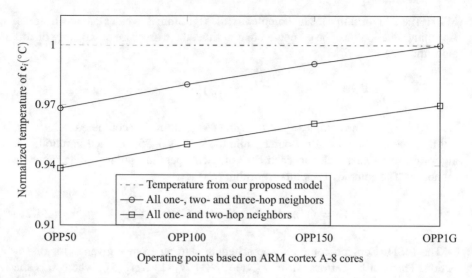

Fig. 4.8 Temperature variation of the proposed model

4.6.2 Validation of the Temperature Model

The temperature model in Eq. (4.15) incorporates only the voltage and frequency of the one-hop neighbors with all other cores operating at the highest operating point of (1.35 V, 1 GHz). To determine the pessimism in this approach, Fig. 4.8 plots the temperature variation obtained using the simplified model of Eq. (4.15), compared to the temperature obtained using the *HotSpot* tool by varying the operating points of the other neighbors. For this experiment, the execution time of the synthetic task is set to 300 s to enable the proposed temperature model to reach its steady-state phase. The thermal data obtained from the *HotSpot* tool are the steady-state values generated by varying the operating point of core c_i and all of its one- and two-hop neighbors in lock-step, with all other cores set as idle. In terms of the *HotSpot* specification, this setup translates to varying the power of c_i and its one- and two-hop neighbors with the values from Table 4.1, and setting the power dissipation as zero for all other cores. The temperature of core c_i obtained from the *HotSpot* tool (in °C) is normalized with respect to the temperature obtained from the model for the different operating points. Similarly, the results for one-, two-, and three-hop neighbors are obtained. As can be seen clearly from the figure, with the one- and two-hop neighboring cores operating at OPP50 (0.93 V, 300 MHz), the temperature from the proposed model is an overestimate by 9.5°C (6.4%). This overestimation decreases as the operating point is increased. This is because, as more cores operate at the highest operating point, the temperature from the model is close to the temperature from the *HotSpot* tool. A similar trend is obtained for the one-, two-, and three-hop neighbors. For this plot, temperature difference between the proposed model and *HotSpot* tool is less than 0.1% at OPP1G.

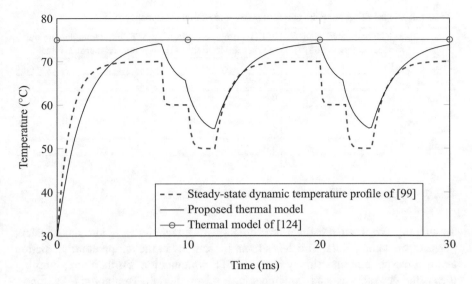

Fig. 4.9 Comparison of temperature results from different thermal models

4.6.3 Comparison with Accurate Temperature Model

Our proposed temperature model is compared with the steady-state dynamic temperature profile (SSDTP) generated using the iterative technique of [14], and the steady-state temperature model of [8]. A synthetic SDFG is considered for this experiment with throughput constraint of 80 iterations per second i.e., a steady-state period of 12.5 ms. This SDFG is executed on a multiprocessor system with 9 cores. The power profile of the SDGF varies within iteration, and this variable power profile is repeated every iteration. The steady-state temperature also varies according to this periodic power pattern as shown clearly in Fig. 4.9, with the red dashed line showing the results obtained using the temperature model of [14]. For the same power profile, our proposed model is shown with black solid line. The mean temperature for this two thermal plots are 63.5°C and 66.1°C, respectively. The temperature model of [8] assumes a steady-state value for the duration of operation, which corresponds to the average power in this duration. This is shown with blue solid line in the figure and corresponds to a temperature of 75°C. (11.5°C difference from the average temperature of [14]). Compared to the accurate thermal model of [14], the proposed thermal model is more accurate than the model of [8].

It is to be noted that, although the proposed model results in an average temperature close to that obtained using the accurate model of [14], the thermal cycling is not captured accurately leading to a misprediction of the thermal cycling related MTTF. However, the advantage is its simpler form (non-iterative as opposed to the iterative technique of [14]), which can be included in the design space exploration, especially for multi-application use-cases.

Table 4.3 Impact of ignoring the transient phase and spatial dependency

Apps	MTTF computed using the thermal model of [9]	MTTF computed using the thermal model of [1]	MTTF computed using our proposed thermal model
FFT	6.1	5.4	6.7
MPEG4	7.2	6.8	8.5
JPEG	8.6	9.4	9.6
MP3	6.4	6.1	7.5
SRC	7.9	8.7	8.7
synth16	6.8	6.0	6.8

4.6.4   Impact of Temperature Misprediction

To simplify the thermal estimation, some existing techniques ignore the transient phase of the temperature. This leads to an inaccuracy in the temperature prediction and a corresponding inaccuracy in the MTTF computation. Furthermore, ignoring the spatial dependency leads to temperature misprediction. To highlight the importance of transient phase and the spatial dependency of the temperature on the MTTF computation, an experiment is conducted with six applications (five real-life and one synthetic) on a multiprocessor platform with nine cores. Table 4.3 reports three MTTF values (in years): the MTTF obtained using the proposed technique with the thermal model proposed in [9] with steady-state temperature phase only; the MTTF obtained using the proposed technique with the temperature model of [1] that considers the temporal dependency only; and the MTTF obtained using the proposed technique with our proposed temperature model.

For the FFT application, MTTF considering our proposed thermal model is 10% and 24% higher as compared to the MTTF considering the thermal model of [9] and [1], respectively. The MTTF improvement by ignoring the spatial dependency (column 3 vs column 4) is higher than the MTTF improvement ignoring the transient phase (column 2 vs column 4). A similar trend is observed for MP3 Decoder and H.264 Encoder. These results clearly demonstrate the importance of spatial component for temperature estimation. A point to note here is that the MTTF improvement by ignoring the spatial dependency is dependent on the size of the application executed on the platform. For JPEG application that uses only two cores of the architecture, the improvement is less than 3%. A similar trend is observed for Sample Rate Converter (SRC) application. Finally, as discussed in Sect. 4.1, considering the steady-state temperature is accurate only if the execution times of the actors of an application are comparable to the thermal time constant of the RC equivalent circuit. This is shown for the synthetic application synth16 (with 16 actors) in the table. The execution times of the actors are generated with a mean of 200 s and standard deviation of 20 s. As can be seen, the MTTF obtained using the proposed model and the model of [9] are the same. On average for all the applications considered, the proposed model improves MTTF by 8% compared to the model in [9], and 15% compared to that of [1].

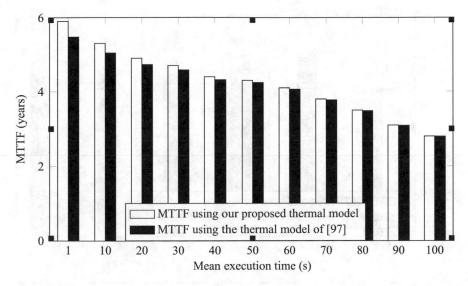

Fig. 4.10 MTTF difference between our thermal model and that using steady-state temperature only

To give further insight into the temperature misprediction considering the steady-state thermal model of [9], experiments are conducted with an SDFG with 16 actors. Actor mapping on the architecture and the schedule of this SDFG are generated using the SDF^3 tool. Next, the ordering of the actors on each core is retained (discarding the timing information), and the average execution times of the actors are varied from 1 to 100 s in steps of 10 s. The MTTF computed using the thermal model of [9] is normalized with respect to the MTTF computed using our proposed thermal model. This is shown in Fig. 4.10. As can be seen clearly from the figure, the MTTF decreases with an increase in the average execution time. This is because, with increase in the average execution time of the actors, stress on the architecture increases, reducing the MTTF. Furthermore, for small mean execution time, the MTTF using our proposed model is higher than that of [9] by 7%. As the mean execution time is increased, the two models become close and temperature difference is less than 0.01%.

Thus far, the validation of the proposed temperature model is presented. In the next few subsections, results to validate the proposed approach are presented.

4.6.5 MTTF Improvement Considering Task Remapping

Modern multiprocessor systems support remapping of tasks (actors in the SDFG terminology) from faults cores to other working cores. MTTF for these systems need to be computed by considering task remapping, as opposed to the naive way

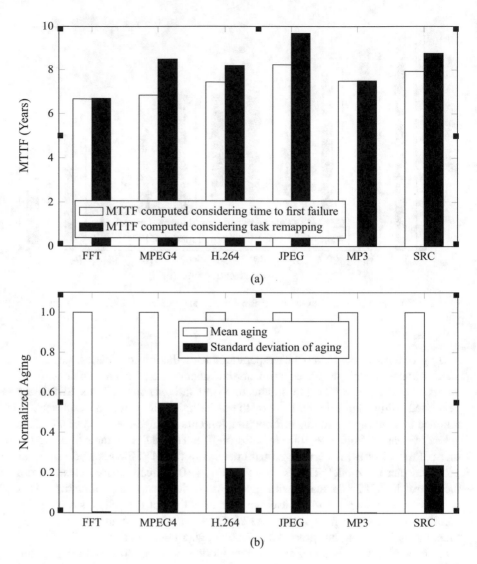

Fig. 4.11 MTTF difference considering time to first failure and task remapping. (**a**) MTTF underestimation. (**b**) Mean and standard deviation of aging

of considering the time to first failure. To highlight the MTTF differences in the two computation techniques, experiments are conducted on a multiprocessor system with six cores and a set of six real-life applications. This is shown in Fig. 4.11a. There are two bars for every application. The left bar is for the MTTF computed considering the first failure and the right one for MTTF computed considering task remapping. As seen from the figure, the two MTTF values are similar for FFT and MP3 decoder applications. For the four other applications, the two MTTF values differ. On average for all applications considered, the MTTF improvement is 15%. To give more insight on the reason for such low MTTF difference for applications such as FFT, as opposed to say, JPEG decoder, Fig. 4.11b plots the mean and the standard deviation of the aging of the different cores for the six applications. The standard deviation of the aging values is a measure of how much the aging of individual cores differ from the mean. A low standard deviation indicates a balanced wear-out of the cores. On the other hand, a high standard deviation indicates some cores age faster than others. The standard deviation is normalized with respect to the mean value of the aging.

As seen from the figure, for applications such as FFT and MP3 Encoder, the standard deviation of the aging is close to zero and thus the wear-out experienced in the cores due to these applications is similar. For these applications, the MTTF computed considering the first failure is similar (difference of 0.5% or lower) to the MTTF computed considering task remapping. This is intuitive, because with all cores suffering similar wear-outs, the breakpoint (the time at which a core fails due to wear-out) for all the cores is similar and therefore remapping leads to an insignificant improvement in lifetime. For all the other applications, the standard deviation is high, with some applications having standard deviation of 60% of the corresponding mean value. For these applications, the aging is not balanced across the cores. Although a balanced aging leads to a higher overall MTTF, a further investigation into these applications reveals that the balanced aging mapping for these applications consumes high energy; therefore, the proposed gradient-based heuristic selects the mapping with non-balanced aging, but with significantly low energy consumption. For these applications, the MTTF computation considering remapping is higher by as much as 24% (average 10%) than the MTTF computation considering the time to the first failure (TTFF).

Finally, Table 4.4 reports the *processor years* (PY) considering the time to first failure (TTFF) and the overall lifetime considering task remapping. For demonstration purpose, only two faults are allowed, and therefore the table reports up to 4 cores used (6 cores in the platform). Column 2 reports the *processor years* considering the time to the first failure. This is the aggregate years spent with all the 6 cores active. The *processor years* with task remapping are shown

Table 4.4 Processor years (PY) considering task remapping

Real-life applications	PY computed considering TTFF	PY computed considering task remapping			
		6 cores	5 cores	4 cores	Total
FFT	37.9	37.9	1.6	0.3	39.8
MPEG4	39.0	39.0	8.8	2.6	50.4
H.264	42.4	42.4	5.5	0.8	48.7
JPEG	46.9	46.9	8.2	2.4	57.5
MP3	42.6	42.6	1.6	0.3	44.5
SRC	45.2	45.2	6.0	0.9	52.1
Average improvement					15.1%

in columns 3, 4, and 5, with the total in column 6. Specifically, column 3 reports the aggregate years spent with all the 6 cores active; column 4 reports the aggregate years with 5 cores active; and column 5 reports the aggregate years spent with 4 cores active. As seen from the table, the total *processor years* considering task remapping for MPEG4 is 30% higher than the *processor years* considering TTFF. This improvement is due to the non-zero *processor years* with 5 and 4 active cores. This improvement demonstrates that, even after the first fault, the multiprocessor system can be exploited to deliver 30% of the performance delivered during the time to the first fault. A similar trend is observed for the other applications in the table. On average, the *processor years* considering task remapping is 15% higher than the *processor years* considering the time to first failure.

4.6.6   Reliability and Energy Improvement

Figure 4.12 plots the energy and MTTF results of our joint optimization technique compared to existing reliability-energy joint optimization technique of [8] for six real-life applications. Additionally, to determine the reliability benefit of the dynamic voltage and frequency scaling, these two techniques are compared with the highest MTTF technique of [4] (referred in the figure as MMax), which determines MTTF by solving a convex optimization problem. These results are represented as three bars corresponding to each application. All results are normalized with respect to the that obtained using the MMax technique.

The following trends can be observed clearly from the figure. Energy consumption using our approach and the existing energy-reliability joint optimization technique of [8] are lower than the highest MTTF technique of [4] that does not consider dynamic voltage and frequency scaling. MTTF computed using these techniques are also higher than the MTTF of [4]. These results signify that, by slowing down of actor computations, reliability of the platform can be improved significantly.

On average for all these applications, the existing joint optimization technique of [8] minimizes energy consumption by 10% and MTTF improvement of 26% compared to the highest MTTF technique of [4]. This technique is, however, based on sequential execution of applications; therefore, the throughput slack (difference between the actual throughput and the throughput constraint) is low, implying a limited scope for actor slowdown. The energy improvement in this technique is, therefore, not significant. Our proposed technique achieves better results than this technique by minimizing energy consumption further by an average 15%, and increasing MTTF by an additional 18%. Compared to [4], our technique minimizes energy consumption by 24% and increases MTTF (lifetime) by 47%. These improvements can be attributed to

- our proposed temperature model that considers transient and steady-state phases as opposed to considering the steady-state temperature only;

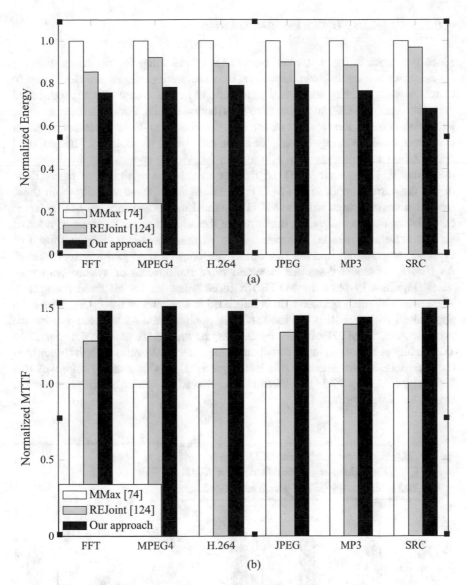

Fig. 4.12 Energy-reliability joint optimization results. For energy, lower is better. For MTTF, higher is better. (**a**) Comparing the total energy consumption of proposed technique with state-of-the-art. (**b**) Comparing the MTTF computed using proposed technique with state-of-the-art

- MTTF computation considering task remapping; and
- pipelined scheduling technique of the proposed approach as opposed to the sequential execution of [8].

4.6.7 Use-Case Optimization Results

Since our work is the first work on MTTF optimization for use-cases, there is no reference for comparison. However, two standard strategies are developed to distribute cores among concurrent applications in a use-case—throughput-based core distribution (TCD) and equal core distribution (ECD). For implementing these strategies, cores of the architecture are first distributed to the applications based on the corresponding strategy (equally or in the ratio of the throughput). The proposed optimization technique is then applied on individual applications to determine their MTTF. The overall MTTF of the use-case is the minimum of the MTTFs of the concurrent applications. MTTFs obtained for a use-case using both these strategies are compared with the MTTF obtained using the proposed MTTF-based core distribution technique. To demonstrate the advantage of our strategy, a set of six synthetic use-cases are generated. Four of these uses-cases are composed of synthetic applications and the two others are composed of real-life applications. As before, these use-cases are executed on a multiprocessor system with nine cores. Figure 4.13 plots the MTTF for three strategies for all these uses-cases. The composition of each use-case is indicated in the label of the figure, where the application with alphabets are the synthetic applications. As an example, for the use-case A-B, the MTTF obtained by distributing the cores equally is 4.6 years. The TCD achieves better results by distributing the cores in the ratio of their throughput requirements. Improvement in this technique is 27%. Our strategy improves this further by achieving 3% higher lifetime. To give insight behind this improvement, a simple example is provided.

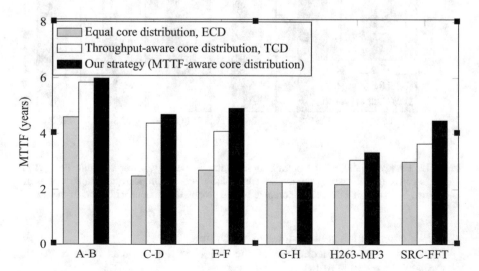

Fig. 4.13 MTTF results with synthetic use-cases for three strategies

Let us consider a multiprocessor system with four cores and a use-case with two applications—synthA and synthB. Furthermore, the throughput requirement of synthA is assumed to be three times more than that of the synthB. The minimum number of cores required to satisfy the throughput requirement of synthA and synthB are two and one, respectively. Let us also assume for the sake of argument, synthB stresses the system more (producing higher temperature) than synthA due to longer execution times of actors in synthB. Distributing the cores to these applications as 3:1 will not be optimal for MTTF. This example motivates and proves the importance of considering MTTF while distributing the cores of an architecture. As seen from the figure, for some use-cases, such as A-B and G-H, improvements using the our strategy are comparable to that of TCD and ECD. For other use-cases, improvements are more than 20%. On average for all these use-cases, our strategy improves MTTF by 10% compared to TCD and 140% compared to ECD.

4.7 Remarks

In this chapter, a simplified temperature model is proposed, based on off-line thermal characterization using the *HotSpot* tool. Based on this model, a gradient-based fast heuristic is proposed to determine voltage and frequency of cores such that the energy consumption is minimized, simultaneously maximizing the system mean time to failure (MTTF). Experiments are conducted on a multiprocessor system using a set of synthetic and real-life application SDFGs, executed individually as well as concurrently. Results demonstrate that the proposed approach minimizes energy consumption by an average 24% and maximizes lifetime by 47% compared to state-of-the-art. Additionally, the proposed MTTF-aware core distribution for concurrent applications results in an average 10% improvement in lifetime compared to the performance-aware core distribution.

References

1. M. Bao, A. Andrei, P. Eles, Z. Peng, Temperature-aware idle time distribution for energy optimization with dynamic voltage scaling, in *Proceedings of the Conference on Design, Automation and Test in Europe (DATE)* (European Design and Automation Association, 2010), pp. 21–26
2. D. Bertozzi, A. Jalabert, S. Murali, R. Tamhankar, S. Stergiou, L. Benini, G. De Micheli, NoC synthesis flow for customized domain specific multiprocessor systems-on-chip. IEEE Trans. Parallel Distrib. Syst. (TPDS) **16**(2), 113–129 (2005)
3. G. Coley, *Beagleboard System Reference Manual* (BeagleBoard.org, 2009), p. 81
4. A. Das, A. Kumar, B. Veeravalli, Reliability-driven task mapping for lifetime extension of networks-on-chip based multiprocessor systems, in *Proceedings of the Conference on Design, Automation and Test in Europe (DATE)* (European Design and Automation Association, 2013), pp. 689–694

5. A. Das, A. Kumar, B. Veeravalli, Energy-aware task mapping and scheduling for reliable embedded computing systems. ACM Trans. Embed. Comput. Syst. (TECS) **13**(2s), 72:1–72:27 (2014)
6. A.S. Hartman, D.E. Thomas, B.H. Meyer, A case for lifetime-aware task mapping in embedded chip multiprocessors, in *Proceedings of the Conference on Hardware/Software Codesign and System Synthesis (CODES+ISSS)* (ACM, 2010), pp. 145–154
7. J. Hu, R. Marculescu, Energy-aware communication and task scheduling for network-on-chip architectures under real-time constraints, in *Proceedings of the Conference on Design, Automation and Test in Europe(DATE)* (IEEE Computer Society, 2004), p. 10234
8. L. Huang, Q. Xu, Energy-efficient task allocation and scheduling for multi-mode MPSoCs under lifetime reliability constraint, in *Proceedings of the Conference on Design, Automation and Test in Europe (DATE)* (European Design and Automation Association, 2010), pp. 1584–1589
9. L. Huang, F. Yuan, Q. Xu, On task allocation and scheduling for lifetime extension of platform-based MPSoC designs. IEEE Trans. Parallel Distrib. Syst. (TPDS) **22**(12), pp. 2088–2099 (2011)
10. A. Leroy, D. Milojevic, D. Verkest, F. Robert, F. Catthoor, Concepts and implementation of spatial division multiplexing for guaranteed throughput in networks-on-chip. IEEE Trans. Comput. **57**(9), 1182–1195 (2008)
11. W. Liao, L. He, K. Lepak, Temperature and supply voltage aware performance and power modeling at microarchitecture level. IEEE Trans. Comput. Aided Des. Integr. Circuits Syst. (TCAD) **24**(7), 1042–1053 (2005)
12. K. Skadron, M.R. Stan, K. Sankaranarayanan, W. Huang, S. Velusamy, D. Tarjan, Temperature-aware microarchitecture: modeling and implementation, ACM Trans. Archit. Code Optim. (TACO) **1**(1), 94–125 (2004)
13. S. Stuijk, M. Geilen, T. Basten, SDF3: SDF for free, in *Proceedings of the International Conference on Application of Concurrency to System Design (ACSD)* (IEEE Computer Society, 2006), pp. 276–278
14. I. Ukhov, M. Bao, P. Eles, Z. Peng, Steady-state dynamic temperature analysis and reliability optimization for embedded multiprocessor systems, in *Proceeding of the Annual Design Automation Conference (DAC)* (ACM, 2012), pp. 197–204
15. W. Zhao, Y. Cao, Predictive technology model for nano-CMOS design exploration. ACM J. Emerg. Technol. Comput. Syst. (JETC) **3**(1), (2007)

Chapter 5
Reliability and Energy-Aware Co-design of Multiprocessor Systems

5.1 Introduction

An emerging trend in multiprocessor design is to integrate reconfigurable area alongside homogeneous processing cores. Hardware–software co-design of these reconfigurable multiprocessor systems needs to address the following two aspects:

Hardware–Software Task Partitioning Given a reconfigurable multiprocessor system and an application represented as a directed graph, the hardware–software task partitioning problem is to determine the tasks of the application that need to be executed on the processing cores (software tasks) and those required to be implemented as hardware on the reconfigurable area (hardware tasks). This problem has been studied extensively in literature to maximize performance and to minimize energy consumption [13].

Hardware Sizing The hardware sizing problem for reconfigurable multiprocessor systems is to determine minimum resources (number of processing cores and the size of reconfigurable area) needed to guarantee performance of every application (enabled individually or concurrently) while satisfying design area budget [7].

Figure 5.1 shows the hardware–software co-design approach for the given set of applications. The approach invokes the hardware–software task partitioning step for every application iteratively, to check if the application performance is met. If this is violated, additional resources are allocated and the hardware–software task partitioning step is repeated; otherwise, the analysis terminates for the application and the process is invoked for the next application. The final platform is determined as the maximum of resources of all applications, enabled individually and concurrently.

Existing studies in this hardware–software co-design suffer from the following limitations. First, checkpointing is used for reconfigurable multiprocessor systems to tolerate transient faults in the processing cores. These solutions guarantee or maximize fault-free task execution on the cores while exploiting execution

© Springer International Publishing AG 2018
A.K. Das et al., *Reliable and Energy Efficient Streaming Multiprocessor Systems*,
Embedded Systems, https://doi.org/10.1007/978-3-319-69374-3_5

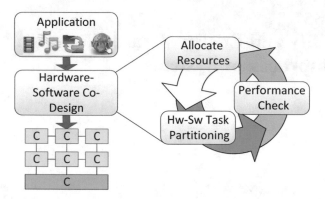

Fig. 5.1 Hardware–software co-design methodology

slack arising from the hardware execution of certain tasks. Area and performance overhead for the fault-tolerance of logic implemented on the reconfigurable area are not accounted in these techniques. Further, extent of fault-tolerance achieved within an allocated reconfigurable area in a co-design framework is not addressed. A complete solution to fault-tolerance needs to incorporate transient fault-tolerance overheads for both software and hardware tasks while satisfying design performance constraints and reconfigurable area availability.

Second, no existing hardware–software task-partitioning techniques consider wear-out of processing cores and transient faults simultaneously. As shown in Sect. 5.2.4, improving transient fault-tolerance by increasing the number of checkpoints negatively impacts lifetime reliability of cores due to wear-out. A balance of the two is essential to mitigate transient faults and wear-out jointly for multiprocessor systems. Moreover, the only existing wear-out aware co-design technique for a static multiprocessor system leaves a significant scope of improvement both in terms of reliability and resource usage when applied to a reconfigurable system.

Third, multimedia applications such as H.264 encoder, decoder, JPEG decoder, etc. are characterized by cyclic dependency of tasks, and require a fixed throughput to be satisfied to guarantee quality-of-service to end users. Existing studies on hardware–software co-design are based on acyclic graph model of applications without considering throughput degradation. These techniques require significant modification (if at all applicable) for streaming multimedia applications represented as synchronous data flow graphs (SDFGs). Last, no existing works consider multi-application use-cases, which is commonly enabled for most multiprocessor systems.

In this work, a design-time technique is proposed for hardware–software partitioning of an application, i.e., determining the tasks to be executed on the processing cores and those on the reconfigurable area. The objective is to improve fault-tolerance of the platform considering three effects—transient faults occurring in the processing cores, single event upsets occurring in the logic configuration bits of the reconfigurable area, and the wear-out of the processing cores. The transient faults in the cores are mitigated using checkpoints.[1] Single event upsets in the logic configuration bits of the reconfigurable area (like Xilinx FPGA) manifest as permanent faults and render the affected logic useless, unless reprogrammed. The proposed approach does not consider reprogramming the reconfigurable area within an application execution; therefore, redundancy-based techniques are used for the single event upsets. The corresponding area overhead is incorporated in the problem formulation. Finally, wear-out of the cores is mitigated using intelligent task mapping and scheduling. Based on the proposed hardware–software task partitioning technique, a hardware–software co-design approach is proposed to determine the minimum resources needed to map and guarantee throughput of applications in all use-cases, simultaneously improving the lifetime reliability measured as mean time to failure (MTTF) and satisfying the specified energy budget and design cost. Following are the key contributions of this chapter.

- Formulation of the wear-out of processing cores and checkpoint-based transient error recovery problem in the hardware–software task partitioning framework with reconfigurable area as a constraint;
- Design space exploration for application partitioning for reconfigurable multiprocessor systems;
- Reliability-aware hardware–software co-design framework incorporating the proposed design space exploration technique;
- Integer linear programming (ILP)-based merging of Pareto-optimal solutions for individual applications to determine the resource requirement for multiapplication use-cases;
- Considering SDFG for hardware–software co-design of multiprocessor system.

The remainder of this chapter is organized as follows. The proposed reliability-aware hardware–software task partitioning technique is discussed in Sect. 5.2. The co-design framework is introduced next in Sect. 5.3. Results are presented in Sect. 5.4 and the chapter is concluded in Sect. 5.5.

5.2   Reliability-Aware Hardware–Software Task Partitioning

In this section, an application partitioning technique is introduced for reconfigurable multiprocessor systems with reliability as an optimization objective. The proposed technique generates the following decisions for every target application.

1. Tasks to be mapped on the processing cores (software tasks) and tasks to be implemented on the reconfigurable area (hardware tasks);

---

[1]Of the different transient fault-tolerant techniques available for processing cores, such as error correction coding, duplication with re-execution and checkpointing as highlighted in Chap. 1, we selected checkpointing as this provides the best trade-off between fault-tolerance and execution/resource overhead.

2. Number of checkpoints for each software tasks; and
3. Mapping and scheduling of software and hardware tasks on the given platform.

These decisions are computed off-line at design-time; mapping and reconfig-urable area configuration (e.g., bit stream) are stored in a database for every application. When an application is enabled at run-time, corresponding mapping and configuration data are fetched. The reconfigurable area is programmed with the bit stream, and task mapping is applied to execute tasks of the application. Dynamic partial reconfiguration will be addressed in future.

5.2.1 Application and Architecture Model

An application is represented as SDFG $\mathscr{G}_{\mathrm{app}} = (\mathbb{A}, \mathscr{C})$, where $\mathbb{A}$ is the set of actors representing tasks of the application and $\mathscr{C}$ is the set of directed edges representing data dependency among various actors. The number of actors is represented as N_a, i.e., $(N_a = |\mathbb{A}|)$. Every actor $\boldsymbol{a}_i \in \mathbb{A}$ is a tuple $\langle \mathrm{RA}_i, t_i, \tau_i \rangle$, where RA_i is the area required to implement $\boldsymbol{a}_i$ on the reconfigurable area, t_i is the time taken by $\boldsymbol{a}_i$ to execute on a dedicated hardware, and τ_i is its execution time on a core. If an actor does not support hardware implementation, value of this parameter is set to infinite. The area overhead for redundancy-based transient fault-tolerance for actor $\boldsymbol{a}_i$ is incorporated into RA_i.

A reference architecture is shown in Fig. 5.2b and consists of N_c homogeneous cores connected to a shared reconfigurable area. The reconfigurable area is a one-dimensional (1D) model and is divided into N_r equal sized frames. A frame is a basic unit for reconfiguration. The architecture is represented as a graph $\mathscr{G}_{\mathrm{arc}} = (\mathbb{C}, \mathbb{E})$, where $\mathbb{C}$ is the set of nodes of the graph and represents cores of the architecture and $\mathbb{E}$ is the set of connections representing links between the cores. The number of cores in the architecture is denoted by N_c, i.e., $N_c = |\mathbb{C}|$. In this work, reconfigurable

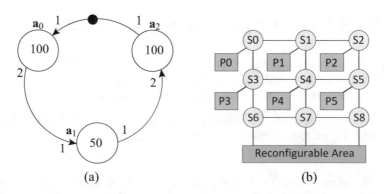

(a)   (b)

Fig. 5.2 Application and architecture model. (**a**) Applications modeled as SDFG. (**b**) An example mesh architecture model

area is considered as a virtual core in the graph. This is because, a reconfigurable area can execute a task when programmed with either the RTL code of a processing core or the hardware implementation of the task. The extended graph with the virtual core is denoted as $\mathcal{G}_{arc}^{N_c+1}$. The following conditions apply for the virtual core.

1. Actors mapped to the virtual core have associated reconfigurable area cost. This is because, an actor mapped on the virtual core implies dedicated hardware for the actor (in reality) that consumes few frames of the reconfigurable area.
2. Multiple independent actors can be executed at the same time on the virtual core. This is because, independent actors implemented on different regions of the reconfigurable area can run independently and concurrently. Although this leads to a contention on the communication link between the processing cores and the reconfigurable area, investigation on the communication aspect of reconfigurable multiprocessor systems is left as future work.
3. An actor mapped on the virtual core does not need software-based protection techniques, such as **checkpointing** and **rollback**. Instead, the protection is provided by replicating the hardware implementation.

This work analyzes the trade-off between transient fault-tolerance and lifetime reliability, considering the impact of temperature from mapping and scheduling decisions. The mapping is specified in terms of the *actor distribution* variable $x_{i,j}$ and the *checkpoint assignment* variable $z_{i,k}$ defined as

$$x_{i,j} = \begin{cases} 1 & \text{if actor } \boldsymbol{a}_i \text{ is assigned to core } \boldsymbol{c}_j \\ 0 & \text{otherwise} \end{cases} \tag{5.1}$$

$$z_{i,k} = \begin{cases} 1 & \text{if actor } \boldsymbol{a}_i \text{ is assigned } k \text{ checkpoints} \\ 0 & \text{otherwise} \end{cases} \tag{5.2}$$

The mapping of $\mathcal{G}_{app}$ on $\mathcal{G}_{arc}$ is represented in terms of two matrices $(\mathcal{M}_d\ \mathcal{M}_c)$, where $\mathcal{M}_d$ and $\mathcal{M}_c$ are defined as

$$\mathcal{M}_d = \begin{pmatrix} x_{0,0} & x_{0,1} & \cdots & x_{0,N_c} \\ x_{1,0} & x_{1,1} & \cdots & x_{1,N_c} \\ \vdots & \vdots & \ddots & \vdots \\ x_{N_a-1,0} & x_{N_a-1,1} & \cdots & x_{N_a-1,N_c} \end{pmatrix} \tag{5.3}$$

$$\mathcal{M}_c = \begin{pmatrix} z_{0,0} & z_{0,1} & \cdots & z_{0,N_c-1} \\ z_{1,0} & z_{1,1} & \cdots & z_{1,N_c-1} \\ \vdots & \vdots & \ddots & \vdots \\ z_{N_a-1,0} & z_{N_a-1,1} & \cdots & z_{N_a-1,N_c-1} \end{pmatrix} \tag{5.4}$$

It is to be noted that the actor distribution matrix $\mathcal{M}_d$ includes the mapping variable for the virtual core (i.e., the RA). Throughout the rest of this chapter, the virtual core is indexed by N_c and the homogeneous cores using indices $0, 1, \ldots,$ $N_c - 1$.

5.2.2 Reliability Modeling Considering Single Actor

Checkpoint refers to the state of the system at a particular instance of time. The interval between two successive checkpoints is called the checkpoint interval. The actor resumes execution at the beginning of each checkpoint interval. When transient faults occur during a checkpoint interval, results generated from that interval are discarded and execution of the actor is repeated from the last valid checkpoint. One important parameter of checkpointing is the checkpoint overhead, defined as the increase in execution time due to the process of checkpointing. This overhead depends on

1. the number of checkpoints, N;
2. the time for checkpoint capture and storage, T_o;
3. the time for recovery from a checkpoint, T_r; and
4. the fault arrival rate, λ.

Following are the assumptions for checkpointing, similar to the works in [1, 5].

- transient faults follow Poisson distribution with a rate of λ failures per unit time;
- transient faults are point failures, i.e., these faults induce errors in the checkpoint interval once and then disappear;
- these faults are statistically independent; and
- checkpoints can be inserted anywhere during an actor execution. This assumption, although difficult to accomplish in practice, simplifies the derivation.

Figure 5.3 shows an example actor execution with N checkpoints. Let T denote the actual computation time of the actor and T_c, the computation time of the actor in each checkpoint interval. Clearly, $T_c = \frac{T}{N+1}$. Assuming Poisson fault arrival, the probability of k faults in the interval t and $t + \Delta t$ is

$$P(t, t + \Delta t, k) = \frac{e^{-\lambda \Delta t} (\lambda \Delta t)^k}{k!} \text{ for } k = 0, 1, \ldots \quad (5.5)$$

Fig. 5.3 Task execution overhead with multiple checkpoints. (**a**) Task execution without checkpoints. (**b**) Task execution with checkpoints

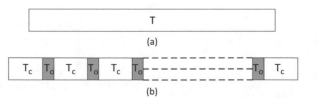

(a)

(b)

Therefore, the probability of at least one fault in the interval Δt is

$$P(t, t + \Delta t, k \geq 1) = 1 - P(t, t + \Delta t, k = 0) = 1 - e^{-\lambda \Delta t} \quad (5.6)$$

Assuming the interval Δt as the checkpoint interval $(T_c + T_o)$, the probability of at least one fault in this interval is $P_e = 1 - e^{-\lambda(T_c + T_o)}$. The expected length of checkpoint interval $E[T_c]$ is calculated as

$$E[T_c] = P\{\text{no fault}\} \cdot \text{original checkpoint interval}$$
$$+ P\{\text{fault}\} \cdot \text{modified checkpoint interval} \quad (5.7)$$

When there are no faults in a checkpoint interval, the duration of this interval is $T_c + T_o$, where T_o is the time for checkpoint computation and storage (refer to Fig. 5.3). Let T_f denote the time of fault from the start of a checkpoint interval. Since faults can occur at any time in the checkpoint interval with a probability P_e, T_f is uniformly distributed in the range 0 to $(T_c + T_o)$ with an average value of $\frac{T_c + T_o}{2}$. Hence, the modified checkpoint interval is given by $T_f + T_r + (T_c + T_o)$, where the first term is the useful computation lost since the beginning of the checkpoint interval, the second term is the time for recovery from the last valid checkpoint, and the last term is the re-execution time of the checkpoint segment starting from the last valid checkpoint. The recovery time includes the overhead to fetch a checkpoint from a local or remote memory and re-loading registers of the faulty core. Equation (5.7) can be written as

$$E[T_c] = (1 - P_e) \cdot (T_c + T_o) + P_e \cdot (\tau + T_r + T_c + T_o)$$
$$= \frac{3(T_c + T_o)}{2} + T_r - \left(\frac{T_c + T_o + 2T_r}{2} \right) e^{-\lambda(T_c + T_o)} \quad (5.8)$$

Expected length of the last checkpoint interval $(E[T_c^L])$ is computed from the above equation by replacing $(T_c + T_o)$ with T_c. This is because there is no checkpoint overhead for the last interval. Overall, the expected execution time of an actor is given by

$$E[T] = N \cdot E[T_c] + E[T_c^L] \quad (5.9)$$

The reliability of an actor a_i with N checkpoints is given by

$$R_i^C(t) = (1 - P_e)^{N+1} + \binom{N+1}{1} P_e (1 - P_e)^{N+1} + \binom{N+2}{2} P_e^2 (1 - P_e)^{N+1} + \cdots$$
$$(5.10)$$

where the first term on the right-hand side is the reliability with no faults, the second term is the reliability with one transient fault in any of the $N+1$ checkpoint intervals, the third term is the reliability with two faults in $N + 2$ intervals ($N + 1$ original intervals and 1 re-execution interval of the interval where the first fault occurs), and so on. Assuming infinite faults in task execution, the above expression reduces to

$$R_i^C(t) = \sum_{\omega=0}^{\infty} \binom{N + \omega}{\omega} P_e^\omega (1 - P_e)^{N+1} = 1 \qquad (5.11)$$

This result is intuitive: if re-execution is allowed every time a fault is detected, an actor will eventually be executed successfully. However, for throughput constrained real-time systems, infinite faults will lead to throughput violation. For these systems, the sum in Eq. (5.11) is evaluated from 0 to ζ_i, where ζ_i is the maximum number of faults that can be tolerated for actor a_i such that its throughput constraint is satisfied. Reliability of actor a_i is

$$R_i^C(t) = \sum_{\omega=0}^{\zeta_i} \binom{N + \omega}{\omega} P_e^\omega (1 - P_e)^{N+1} \qquad (5.12)$$

Figure 5.4a,b plots the reliability of an actor with execution times of $T = 50$ ns and $T = 150$ ns, respectively as the number of checkpoints is increased from 0 to 20. These results are for ζ values of 5 and 10. As seen from these figures, the reliability first increases with the number of checkpoints. After it reaches maximum, the reliability decreases with increase in the number of checkpoints. This trend is consistent to that discussed in [8, 9]. Moreover, reliability increases with an increase in the number of faults that can be tolerated, i.e., ζ (consistent with Eq. (5.12)). Finally, the reliability, corresponding to a particular number of checkpoints and ζ value, decreases as the actor execution time is increased from 50 to 150 ns. This is because, with increase in the execution time, length of the checkpoint interval increases ($T_c = \frac{T}{N+1}$) and so does the fault probability P_e (Eq. (5.6)). This reduces the reliability.

Figure 5.4c,d plots the dependency of checkpoint overhead (T_o) on the highest reliability value. The checkpoint overhead is expressed as a percentage of the execution time in these figures. As seen from these figures, the highest reliability point decreases with an increase in the overhead. This is because, with increase in the checkpoint overhead, the time between two successive checkpoints ($T_c + T_o$) increases, leading to an increase in fault probability.

5.2.3 Reliability Modeling with Multiple Interconnected Actors

Lifetime Reliability The MTTF of a system considering an application with multiple interconnected actors is given by the *Max-Min* approach (Chap. 2). Thus,

$$\text{MTTF} = \min_{j}\{\text{MTTF}_j\} \qquad (5.13)$$

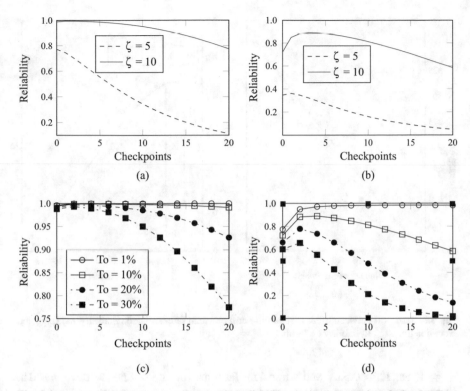

Fig. 5.4 Impact of different parameters on the reliability considering transient faults. (a) $T = 50\,\text{ns}$. (b) $T = 150\,\text{ns}$. (c) $T = 50\,\text{ns}$. (d) $T = 150\,\text{ns}$

Reliability Considering Checkpoints For an application consisting of multiple interconnected actors, the overall reliability considering transient faults is calculated based on the assumption that (1) transient faults are independent, and (2) an application is successful when all actors of the application execute successfully. Reliability and mean time between failures (MTBF) are given by

$$R_T(t) = \prod_{i=1}^{N_a} R_i^C(t) \text{ and MTBF} = \int_{t=0}^{\infty} R_T(t)dt \qquad (5.14)$$

where N_a is the number of actors and $R_i^C(t)$ is the reliability of actor a_i.

5.2.4   Lifetime Reliability and Transient Fault Reliability Trade-Off

Figure 5.5a plots the expected execution time of an actor as the number of checkpoints is increased. Parameters used for this simulation are as follows: $T = 150\,\text{ns}$,

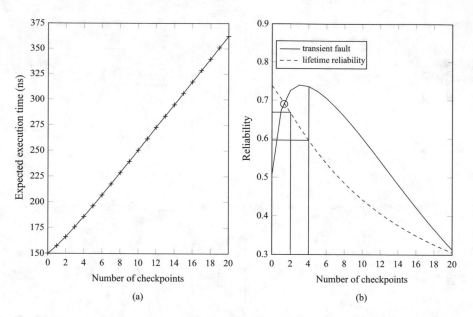

Fig. 5.5 Trade-off between lifetime reliability and transient faults related reliability. (**a**) Expected execution time. (**b**) Reliability trade-off

$T_o = 15$ ns, $T_r = 0.5$ ns, and a transient fault arrival rate of one fault every 100 h. As seen from the figure, the expected execution time increases with an increase in the number of checkpoints. With every extra checkpoint, the checkpoint interval reduces and therefore, the loss in computation reduces when a fault is detected. However, with every added checkpoint, the corresponding checkpoint overhead is to be included in the execution time. This increases the expected execution time (Eq. (5.9)).

As shown in Chap. 2, the lifetime reliability of a core due to wear-out is negatively dependent on the execution time of the actors executed on the core. Figure 5.5b plots the decrease in lifetime reliability (considering wear-out) with increase in the number of the checkpoints. This is consistent with Eq. (2.12), and is shown in the figure with the dashed line. To show the trade-off between lifetime reliability and the reliability considering transient faults, Eq. (5.12) is also plotted on the same figure, and is shown with the solid line.

As seen from this figure, the two reliability plots intersect when the number of checkpoints = 2. This intersection point depends on the execution time, the number of checkpoints, and the overheads T_o and T_r. Since transient fault reliability first increases and then decreases, there can be potentially two intersection points. Since the number of checkpoints needs to be an integer, the selected values are rounded to the nearest integer. As an example, the cross-over point corresponds to 1.8 checkpoints, which is rounded to 2. Similarly, the number of checkpoints corresponding to the highest transient fault reliability is 3.6 and is rounded to 4.

There are two lines drawn in the figure corresponding to 2 and 4 checkpoints. As seen from this figure, selecting 4 checkpoints (corresponding to the highest transient fault reliability) results in a reduction of the lifetime reliability by 15% as compared to selecting 2 checkpoints (corresponding to the intersection point). This clearly demonstrates the significance of selecting the intersection point for a given application.

5.2.5   Hardware–Software Partitioning Flow

For the hardware–software actor partitioning, a joint metric *reliability gradient* (*rg*) is defined as

$$rg = \frac{\Delta R_P(t)}{\Delta R_T(t)} \tag{5.15}$$

where $R_T(t)$ and $R_P(t)$ are reliabilities considering transient faults and wear-out, respectively. The reliability gradient is interpreted as ratio of the change (decrease) in reliability due to wear-out per unit change (increase) in reliability due to transient faults. The optimization objective is to minimize the reliability gradient. A fast design space exploration technique is proposed. This is shown as pseudo-code in Algorithm 6. A list (*HList*) is defined to store the actors that are to be implemented on the virtual core. The algorithm iterates (lines 2–10) as long as the available reconfigurable area constraint is satisfied (line 2). At every iteration, the algorithm selects one actor from the set $\mathbb{A} \setminus HList$ (i.e., selects one of those actors not marked for virtual core) and assigns it temporarily to the reconfigurable area (line 4). The hardware actors (from *HList*) are mapped on the reconfigurable area and the software actors on the processing cores, and the whole application graph is scheduled

---

Algorithm 6 *HSAP*(): hardware–software actor partitioning

Require: $\mathscr{G}_{app}, \mathscr{G}_{arc}^{N_c+1}$, throughput constraint $\mathbb{T}_c$, size of reconfigurable area S_{RA}
Ensure: $\left(\mathscr{M}_d \ \mathscr{M}_c\right)$
 1: Initialize $HList = \emptyset$
 2: **while** $\sum_{\forall a_k \in HList} RA_k \leq S_{RA}$ **do**
 3:    **for all** $a_i \in \mathbb{A} \setminus HList$ **do**
 4:       $HList.push(a_i)$
 5:       $[\mathscr{M}_d \ \mathscr{M}_c \ rg^i] = FindMinRG(\mathscr{G}_{app}, \mathscr{G}_{arc}^{N_c+1}, \mathbb{T}_c, HList)$
 6:       $HList.pop(a_i)$
 7:    **end for**
 8:    Find $a_j \in \mathscr{G}_{app} \setminus HList$ such that rg^j is minimum
 9:    $HList.push(a_j)$
10: **end while**
11: $[\mathscr{M}_d \ \mathscr{M}_c \ rg] = FindMinRG(\mathscr{G}_{app}, \mathscr{G}_{arc}^{N_c+1}, \mathbb{T}_c, HList)$
12: Return $\left(\mathscr{M}_d \ \mathscr{M}_c\right)$

Algorithm 7 *FindMinRG*(): mapping and scheduling to find the minimum reliability gradient

---

Require: $\mathcal{G}_{app}, \mathcal{G}_{arc}^{N_c+1}$, throughput constraint $\mathbb{T}_c$ and *HList*
Ensure: Mapping $(\mathcal{M}_d\ \mathcal{M}_c)$ and minimum reliability gradient rg
 1: $[\mathcal{M}_d\ \mathcal{S}] = REOpt(\mathcal{G}_{app}, \mathcal{G}_{arc}^{N_c+1}, \mathbb{T}_c, 1)$ //*Wear-out minimum mapping.*
 2: Initialize $\mathcal{M}_c$ with $z_{i,k} = 0\ \forall i, k$ //*Set checkpoints for all actors equal to zero.*
 3: $[R_P^{init}\ R_T^{init}] = CalculateReliability(\mathcal{S}, \mathcal{M}_d, \mathcal{M}_c)$ //*Calculate the initial wear-out and transient fault related reliability using Eqs. (2.12) and (5.14).*
 4: Initialize *runIter* $= 1$ //*Initialize the loop variable.*
 5: **while** *runIter* > 0 **do**
 6:      $[\mathcal{S}\ \mathbb{T}] = MSDF3(\mathcal{M}_d, \mathcal{G}_{app}, \mathcal{G}_{arc}^{N_c+1})$ //*Get the schedule and throughput of the allocation matrix* $\mathcal{M}_d$.
 7:      $[R_P\ R_T] = CalcReliability(\mathcal{S}, \mathcal{M}_d, \mathcal{M}_c)$ //*Calculate reliabilities.*
 8:      $i_{best} = -1; j_{best} = -1; k_{best} = -1; rg_{best} = \infty$ //*Initialize.*
 9:      **for all** $a_i \in \mathbb{A} \setminus HList$ **do**
10:          **for all** $c_j \in \mathcal{G}_{arc}^{N_c+1}$ **do**
11:              **for** $k = 1$ to N_{cp} **do**
12:                  $\left(\mathcal{M}_d^{temp}\ \mathcal{M}_c^{temp}\right) = \left(\mathcal{M}_d\ \mathcal{M}_c\right)$ //*Use temporary allocation and assignment matrices.*
13:                  Update $\left(\mathcal{M}_d^{temp}\ \mathcal{M}_c^{temp}\right)$ with $x_{i,j} = z_{i,k} = 1$ //*Update these matrices.*
14:                  $[\mathcal{S}\ \mathbb{T}] = MSDF3(\mathcal{M}_d^{temp}, \mathcal{G}_{app}, \mathcal{G}_{app}^{N_c+1})$ //*Calculate the new schedule and throughput.*
15:                  $[R_P^{temp}\ R_T^{temp}] = CalcReliability(\mathcal{S}, \mathcal{M}_d^{temp}, \mathcal{M}_c^{temp})$
16:                  $rg = \frac{R_P - R_P^{temp}}{R_T^{temp} - R_T}$ //*Calculate the reliability gradient.*
17:                  **if** $rg < rg_{best}$ && $\mathbb{T} \geq \mathbb{T}_c$ **then**
18:                      $i_{best} = i; j_{best} = j; k_{best} = k; rg_{best} = rg$
19:                  **end if**
20:              **end for**
21:          **end for**
22:      **end for**
23:      **if** $i_{best} \geq 0$ **then**
24:          Update $\left(\mathcal{M}_d\ \mathcal{M}_c\right)$ with $x_{i_{best}, j_{best}} = z_{i_{best}, k_{best}} = 1$
25:      **else**
26:          *runIter* $= 0$ //*No possible remapping w/o violating the throughput constraint.*
27:      **end if**
28: **end while**
29: $[\mathcal{S}\ \mathbb{T}] = MSDF3(\mathcal{M}_d, \mathcal{G}_{app}, \mathcal{G}_{app}^{N_c+1})$ //*Schedule and throughput of the final allocation.*
30: $[R_P^{finl}\ R_T^{finl}] = CalculateReliability(\mathcal{S}, \mathcal{M}_d, \mathcal{M}_c)$ //*Final reliabilities.*
31: $rg = \frac{R_P^{init} - R_P^{finl}}{R_T^{finl} - R_T^{init}}$ //*Overall reliability gradient.*
32: Return $[\mathcal{M}_d\ \mathcal{M}_c\ rg]$

---

to determine the minimum reliability gradient. This step is performed using the *FindMinRG* routine that outputs the actor distribution, checkpoint assignment, and the reliability gradient (line 5). The actor, whose assignment to the reconfigurable area leads to the minimum reliability gradient, is pushed to the *HList* (line 9).

A key component of this algorithm is the *FindMinRG* subroutine, whose pseudo-code is shown in Algorithm 7. The first step of Algorithm 7 is the initial actor distribution that minimizes wear-out (line 1). The algorithm continues to remap ever actor to a core with different checkpoints to determine the reliability gradient (lines 9–16). If the reliability gradient obtained for an assignment is lower than the best value obtained thus far, the best values are updated (lines 17–19). After iterating for all the actors, the actor distribution is changed with the best value of actor, core, and checkpoints. This process is re-iterated (starting with this changed distribution) as long as no further re-mapping is possible without violating the throughput constraint. When this happens, the best value of actor, core, and checkpoints are all negative. The algorithm proceeds to the **else** section (lines 25–27) where the terminating condition is asserted. The final actor distribution and checkpoint assignment matrices $\left(\mathcal{M}_d \ \mathcal{M}_c \right)$ are returned along with the reliability gradient. The actor distribution matrix is used in the MSDF^3 tool (refer to Chap. 4) to generate the throughput and schedule.

The schedule and the mapping matrices are used in *CalcReliability* routine to compute the different reliability values. Specifically, the schedule is used to compute the lifetime reliability, R_P^{temp}, considering the thermal impact as detailed in Chap. 4; the mapping matrices are used to compute the reliability considering checkpointing for transient fault-tolerance, R_T^{temp}, as explained in Sects. 5.2.2 and 5.2.3. Finally, the wear-out minimum initial mapping is the optimization technique proposed in Chap. 4.

5.3 Reliability-Aware Co-design

Figure 5.6 shows the reliability-aware hardware–software co-design approach proposed in this work. The approach consists of two components—reliability-aware design space exploration (*RDSE*) with individual applications and reliability-aware Pareto merging for use-cases. The *RDSE* generates a set of Pareto-points (actor distribution and checkpoint assignment), which are optimal in terms of reliability and resource usage (i.e., the number of processing cores and size of the reconfigurable area). Next, analysis is performed with the given set of use-cases. Specifically, an ILP-based technique is proposed to merge the Pareto-points (obtained in the previous step) of the applications constituting a use-case to determine the minimum resource usage while satisfying the given energy and performance budget.

5.3.1 Design Metrics

Performance The minimum throughput required for an SDFG is denoted by $\mathbb{T}_c$. This is the performance parameter used in the approach and is considered a design constraint.

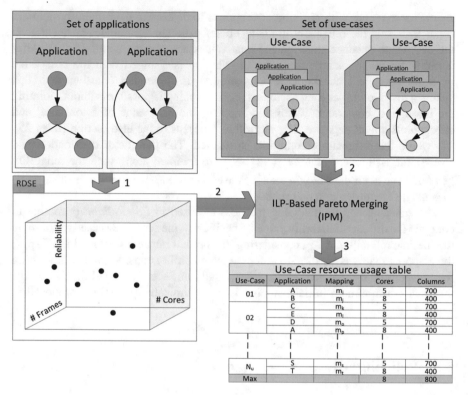

Fig. 5.6 Proposed hardware–software co-design flow

Energy The energy consumption of the multiprocessor system is modeled in Chap. 4. The energy metric is used as a design constraint. Specifically, $E^{tot} \leq E^{max}$, where E^{max} is the given energy budget.

Cost The design cost is specified in terms of the number of cores and the size of the reconfigurable area. The maximum number of cores allowed is denoted by N_c^{max} and the maximum size of the reconfigurable area is denoted by RA^{max}. This is used as a design constraint.

Reliability The reliability considering wear-out and transient faults is modeled in Eqs. (2.12) and (5.14), respectively. The co-design problem is demonstrated here to maximize the lifetime reliability (measured as mean time to failure) under a given constraint of transient fault-tolerance, specified as mean time between failure. However, the technique can be easily adopted to optimize for these parameters (MTTF or MTBF) individually, or combined together in the form of a joint metric.

Algorithm 8 *RDSE*(): design space exploration for reliability-resource trade-off

---

Require: $\mathcal{G}_{app}, \mathcal{G}_{arc}, \mathbb{T}_c, E^{max}, N_c^{max}, RA^{max}$
Ensure: Mapping Pareto database *ParetoDB*
1: **for** $N_c \in [1..N_c^{max}]$ **do**
2:     **for** $\{S_{RA} = e; S_{RA} \leq RA^{max}; S_{RA} = S_{RA} + e\}$ **do**
3:         $(\mathcal{M}_d\ \mathcal{M}_c) = HSAP(\mathcal{G}_{app}, \mathcal{G}_{arc}^{N_c+1}, \mathbb{T}_c, S_{RA})$ *//Hardware–software actor partitioning.*
4:         $E^{tot} = CalculateTotalEnergy(\mathcal{M}_d)$ *//Calculate the total energy of an application.*
5:         $[\mathcal{S}\ \mathbb{T}] = MSDF^3(\mathcal{M}_d, \mathcal{G}_{app}, \mathcal{G}_{arc}^{N_c+1})$ *//Calculate the schedule and throughput.*
6:         $M = CalculateMTTF(\mathcal{S}, \mathcal{M}_d, \mathcal{M}_c)$ *//Calculate the MTTF using Eq. (5.13).*
7:         **if** $\mathbb{T} \geq \mathbb{T}_c$ && $E^{tot} \leq E^{max}$ **then**
8:             $ParetoDB(N_c, S_{RA}).push(\mathcal{M}_d, \mathcal{M}_c, M_J, E^{tot})$ *//Store in a database.*
9:         **end if**
10:     **end for**
11: **end for**
12: Process *ParetoDB* to retain only Pareto-optimal points.

5.3.2   Reliability-Resource Usage Trade-Off

Algorithm 8 provides the pseudo-code of the reliability-aware design space exploration to determine the reliability-resource trade-off. As shown in Sect. 5.1 (refer to Fig. 5.1), the given problem can be solved in a hierarchical manner with the reliability-aware hardware–software actor partitioning on a given reconfigurable multiprocessor system (i.e., with N_c cores where $N_c \leq N_c^{max}$ and *RA* frames where $RA \leq RA^{max}$) at the lower hierarchy, followed by its integration in the higher level problem of determining the reliability-resource trade-off. An epoch ($e = 100$) is defined and the frames allocated to the multiprocessor system are incremented by e at each iteration. The algorithm inputs the application graph and the design constraints—the throughput constraint $\mathbb{T}_c$, the energy budget E^{max}, and the design cost N_c^{max} and RA^{max}. The algorithm then determines the actor distribution and the MTTF value for every combination of core count and reconfigurable area frames by invoking the *HSAP* subroutine (Algorithm 6). The energy and the schedule are computed (lines 4–5). The schedule and the actor distribution are used to compute the MTTF (Eq. (5.13)). If the energy is lower than the energy budget and the throughput constraint is satisfied, the actor distribution, the MTTF, and the energy value are stored in the Pareto database *ParetoDB* corresponding to the number of cores and size of the reconfigurable area (lines 7–9). Finally, the *ParetoDB* is processed to retain only the Pareto-optimal points.

5.3.3   ILP-Based Pareto Merging

The objective in the next stage of the co-design is to determine the reliability (MTTF) maximum mappings for applications enabled simultaneously. These mappings determine the size of the platform. The problem is formulated as a binary integer linear programming (ILP) problem and is solved using Matlab. The inputs to this step are the set of mappings (from *ParetoDB*) for every application of a given use-case. To limit the scalability of this approach (the ILP), only the Pareto-optimal points (i.e., the actor distribution and checkpoint assignments that are optimum with respect to reliability, energy, number of cores, and size of reconfigurable area) are used. Every Pareto-point of an application is associated with four parameters—number of cores, reconfigurable area size, reliability, and energy. This step selects one Pareto-point for every application of a use-case such that the reliability is maximized while satisfying the design cost and energy budgets. To aid the understanding of this approach, the ILP is formulated for a two-application use-case. The technique can be trivially extended to include any multi-application use-cases.

Let *App_A* and *App_B* be the two applications constituting a use-case. The following notations are defined for the ease of problem formulation.

$$N_X = \text{number of Pareto-points of } App\_X \text{ where X = A or B}$$

$$m_1^X, \ldots, m_{N_X}^X = N_X \text{ Pareto-points of } App\_X$$

$$n_i^X = \text{number of cores of Pareto-point } m_i^X \text{ of application } App\_X$$

$$c_i^X = \text{reconfigurable area usage of Pareto-point } m_i^X$$

$$r_i^X = \text{reliability of Pareto-point } m_i^X$$

$$e_i^X = \text{energy consumption of Pareto-point } m_i^X$$

Let Y_{ij} be defined as follows:

$$Y_{ij} = \begin{cases} 1 & \text{if } m_i^A \text{ and } m_j^B \text{ are selected} \\ 0 & \text{otherwise} \end{cases} \tag{5.16}$$

5.3.3.1 Constraints of the ILP

- One Pareto-point for each application is to be selected, i.e., $\sum_{i=1}^{N_A} \sum_{j=1}^{N_B} Y_{ij} = 1$
- Design cost is to be satisfied, i.e.,

$$\sum_{i=1}^{N_A} \sum_{j=1}^{N_B} Y_{ij} \cdot (n_i^A + n_i^B) \leq N_c^{\max} \text{ and } \sum_{i=1}^{N_A} \sum_{j=1}^{N_B} Y_{ij} \cdot (c_i^A + c_i^B) \leq RA^{\max}$$

- Energy budget is to be satisfied, i.e., $\sum_{i=1}^{N_A} \sum_{j=1}^{N_B} Y_{ij} \cdot (e_i^A + e_i^B) \leq E^{\max}$

5.3.3.2   Objective of the ILP

Reliability of the platform with applications A and B executing simultaneously is

$$R_A = \text{Reliability due to } A = \sum_{i=1}^{N_A} \sum_{j=1}^{N_B} Y_{ij} \cdot r_i^A \qquad (5.17)$$

$$R_B = \text{Reliability due to } B = \sum_{i=1}^{N_A} \sum_{j=1}^{N_B} Y_{ij} \cdot r_i^B$$

The overall reliability optimization objective is to maximize $\min\{R_A, R_B\}$.

5.3.3.3   ILP Solution

Let the solution of the ILP be represented as $Y_{ij} = 1$ for $i = i_b$ and $j = j_b$, and $Y_{ij} = 0$ otherwise. The resources used for this use-case (with applications A and B) are as follows: number of cores $= (n_{i_b}^A + n_{j_b}^B)$ and reconfigurable area size $= (c_{i_b}^A + c_{j_b}^B)$. Thus, the ILP is solved for all use-cases to fill the use-case resource table of Fig. 5.6. A point to be noted is that the execution time to fill the use-case resource table is strongly dependent on the number of use-cases. As established in [7], the number of use-cases grows exponentially with the number of applications. Use-case pruning techniques proposed in [7] are adopted here. Once the use-case resource table is filled, the final platform is determined as the highest number of cores and reconfigurable area size.

5.4   Results and Discussions

Fifty synthetic SDFGs are generated from [12] with number of actors between 8 and 32. Additionally, a set of real-life applications (streaming and non-streaming) are considered from [2, 12]. These applications are *H.263 Encoder/Decoder, H.264 Encoder, MP3 Decoder, MPEG4 Decoder, JPEG Decoder, Sample Rate Converter, FFT, iFFT,* and *Romberg Integration.* The hardware–software task partitioning experiments (Sects. 5.4.2–5.4.4) are conducted on a multiprocessor system with 9 cores and 500 frames of reconfigurable area.

5.4.1   Algorithm Complexity

The complexity of Algorithm 6 is computed as follows. Let η be the average number of actors that can be accommodated in the given reconfigurable area. There are

η iterations of the outer while loop (lines 2–10). At each iteration, lines 4–6 are executed for all actors in $\mathbb{A} \setminus HList$. This can be upper bounded by N_a, the total number of actors in the application graph. Line 8 finds the maximum from a list of N_a elements. The complexity of Algorithm 6 is given by

$$C_6 = O\left(\eta \cdot (N_a \cdot O(FindMinRG) + N_a)\right) = O\left(N_a^2 \cdot C_7\right) \qquad (5.18)$$

where $O(FingMinRG) = C_7$ is the complexity of $FindMinRG$ and $\eta \leq N_a$. The complexity of $FindMinRG$ routine is computed as follows. MSDF[3] computes the schedule starting from a given actor allocation. This can be performed in $O\left(N_a \log N_a + N_a \cdot \pounds\right)$ [4] where $\pounds$ is the average number of successors of an actor. The reliability gradient can be computed in $O\left(N_a + N_c\right)$. Assuming the outer while loop executes for χ times on average, the complexity of Algorithm 7 is

$$C_7 = O\left(\chi \cdot N_a \cdot N_c \cdot N_{cp} \cdot (N_a \log N_a + N_a \cdot \pounds + N_a + N_c)\right) = O\left(N_a^4 \cdot N_{cp}\right) \qquad (5.19)$$

using $\pounds \leq N_a$ and $N_c \leq N_a$ and N_{cp} is the maximum number of checkpoints per actor. Combining Eqs. (5.18) and (5.19), the complexity of Algorithm 6 is given by

$$C_6 = O\left(N_a^6 \cdot N_{cp}\right) \qquad (5.20)$$

5.4.2   Reliability Trade-Off Results

Figure 5.7 plots the mean time to permanent failures (MTTF) and mean time between transient faults (MTBF) for different design solutions (actor distribution and checkpoint assignment), outcome of the proposed *HSAP* algorithm for H.263 Decoder application. The figure also plots the solution obtained using the wear-out-aware task mapping technique of [6] (marked in the figure by the alphabet A) and the checkpointing based transient fault-tolerant technique of [10] (marked in the figure by the alphabet B). The transient fault arrival rate of 1 fault every 100 h of operation is considered in this experiment. Reliability requirements are set as follows: MTTF = 3 years and MTBF = 200 h, and are shown in the figure by the solid lines.

As discussed previously, wear-out aware task mapping technique of [6] does not consider transient fault-tolerance, and therefore the MTBF reported in the figure is obtained by considering zero checkpoints in the actor execution. On the other hand, the checkpoint-based transient fault-tolerance technique of [10] does not consider lifetime reliability. The number of checkpoints for the actors is selected such that the reliability is maximized (refer to Fig. 5.5). The proposed *HSAP* algorithm selects the point marked C in the figure. This point satisfies the reliability requirements for both fault types and results in minimum reliability gradient, i.e. the minimum

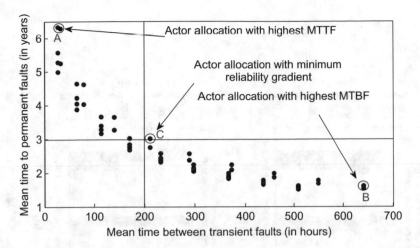

Fig. 5.7 Mean time to failure for transient and permanent faults for H.263 decoder

degradation of lifetime reliability considering wear-out with a maximum increase in the reliability considering transient faults. The actor distribution corresponding to this point improves MTTF by 100% ($2\times$ improvement) compared to [10].

To give more insight into the different trade-off trends obtained for different applications, Fig. 5.8 plots the trade-off between the two reliabilities for four different applications. Results obtained using our proposed technique are compared with [6] and [10]. MTTF and the MTBF constraint for these applications are set to 3 years and 20 h, respectively. The plot for the JPEG decoder (Fig. 5.8b) shows that the MTTF and the MTBF requirements are satisfied by all three techniques. However, the proposed technique and the technique of [10] offer better reliability trade-offs than [6]. Results for this application show that the improvement in the proposed approach is insignificant -18% higher MTTF and 10% lower MTBF compared to [10]. Therefore, both the proposed approach and that of [10] are acceptable solutions, and either of them can be selected as the final solution.

The iFFT (Fig. 5.8c) demonstrates similar trend as H.263 decoder. Out of the three techniques, only the proposed technique satisfies both the MTTF and MTBF requirements. Finally, for the application of MP3 decoder (Fig. 5.8c), the MTTF requirement is violated by all the techniques. This is due to the high stress introduced in the system due to the large number of actors. The MTBF requirement is violated by [6], but both the proposed and the [10] satisfy the MTBF requirement. For this application, our proposed technique results in 125% higher lifetime with less than 15% lower MTBF as compared to [10], clearly demonstrating its superiority. Out of all 60 applications (real-life and synthetic), 6 applications (2 real-life and 4 synthetic) show trends similar to the JPEG decoder and other 3 applications (1 real-life and 2 synthetic) that of the MP3 decoder. Remaining 51 applications (7 real-life and 44 synthetic) show trends similar to H.263 decoder.

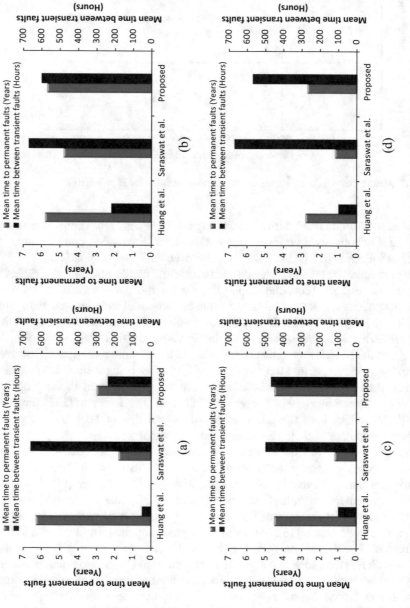

Fig. 5.8 Reliability trade-off results for four different applications. (**a**) H.263 decoder. (**b**) JPEG decoder. (**c**) iFFT. (**d**) MP3 decoder

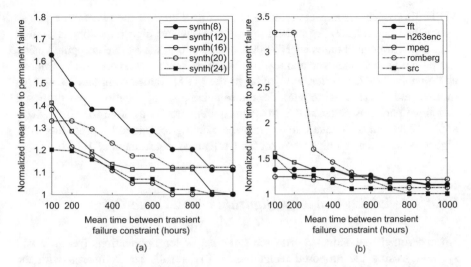

Fig. 5.9 Normalized MTTF with varying transient fault-tolerance constraint [3]. (**a**) Synthetic applications. (**b**) Real applications

To summarize the results for all these sixty applications, the proposed technique improves the platform lifetime by 18–225% (with an average of 60%) compared to the highest MTBF technique of [10]. Finally, the MTTF using the proposed technique is within an average 15% of the highest MTTF of [6].

5.4.3 MTTF Results with Varying MTBF Constraint

Figure 5.9 plots the normalized lifetime (i.e., MTTF) computed using our proposed technique by varying transient fault-tolerance constraint for five synthetic and five real-life applications. Synthetic applications are labeled in the figure with *synth(n)* where *n* denotes the number of tasks of the application; the five real-life applications are *FFT, H.263 Encoder, MPEG4 Decoder, Romberg Integration*, and *Sample Rate Converter*. MTTF computed using the proposed technique is normalized with respect to that obtained using the transient fault-tolerant technique of [10]. The MTBF is varied from 100 to 1000 h. The MTBF constraint is interpreted as follows: MTBF of 100 h implies that with the given fault arrival rate, the system should be capable of correcting these faults using checkpointing mechanism, with the time allowed between two non-correctable faults to be 100 h. Clearly, higher the MTBF requirement, more stringent is the transient fault-tolerance constraint. The range for the MTBF constraint (100–1000 h) represents the varying reliability requirement of both safety and non-safety critical applications.

It can be seen from the figure that, as the transient fault-tolerance constraint becomes more and more stringent (higher MTBF requirement), the normalized

MTTF drops. This is expected due to conflicting nature of the two fault types as established in Sect. 5.2. For critical applications, where high reliability is desired (e.g., non-correctable faults every 1000 h of operation), the MTTF computed using the proposed technique for applications, such as *synth(12)*, *synth(16)*, *synth(24)*, and *Sample Rate Converter*, are similar to the MTTF values computed using [10] (normalized MTTF close to one). For the remaining six applications, our proposed technique performs better at this MTBF requirement. On average for all the sixty applications, our technique outperforms the existing technique by providing 10% higher MTTF even at a high transient fault-tolerance requirement of 1000 h.

5.4.4 Reliability and Reconfigurable Area Trade-Off

An experiment is conducted with the same set of ten applications (real-life and synthetic) using the proposed technique. MTTF results are compared with the MTTF obtained with no reconfigurable area, i.e., on a static multiprocessor system with the same number of cores. Figure 5.10 plots the normalized MTTF as the size of the reconfigurable area is increased. The MTTF obtained using the proposed *HSAP* algorithm for an application is normalized with respect to the MTTF obtained without reconfigurable area. As can be seen from the figure, the MTTF improves with an increase in the size of the reconfigurable area. With increase in the size of the reconfigurable area, more actors can be implemented on hardware (the virtual core), reducing the stress on the processing cores leading to an increase in MTTF.

It is to be concluded from the figure that different applications exhibit different trade-offs with respect to reconfigurable area and lifetime performance. It is

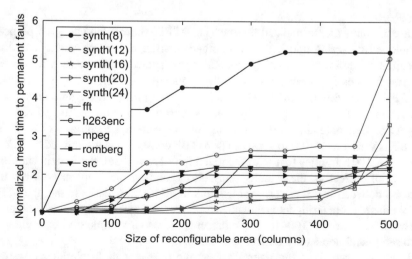

Fig. 5.10 MTTF and reconfigurable area trade-off [3]

essential to characterize each application during the design phase to explore such trade-off. This knowledge can be applied at run-time during application mapping and reconfigurable area distribution among multiple simultaneous applications. As an example, if the reconfigurable area available at a given time during operation is 100 frames and application *FFT* and *Sample Rate Converter* needs to be mapped, it is beneficial to reserve the reconfigurable area for *Sample Rate Converter* that provides significant improvement in lifetime than *FFT*.

5.4.5 *Hardware–Software Co-design Results*

To establish the area and energy overhead introduced for reliability aware co-design, the proposed technique is compared with energy-aware co-synthesis technique of [11] (referred to as *ECosynth*) and the reliability-aware platform generation technique of [14] (referred to as *RPGen*). It is to be noted that the size of platform for *ECosynth* is determined considering energy only, however the platform reliability value is determined using the *HSAP* algorithm proposed in Sect. 5.2. Table 5.1 reports the resource usage (in terms of the number of cores and reconfigurable area), normalized energy and reliability of the proposed technique in comparison with *ECosynth* and *RPGen* techniques for single applications as well as for use-cases. The design cost for the reconfigurable multiprocessor system is set to a maximum of 16 cores and 1000 reconfigurable area columns. The reference for reliability is obtained by mapping an application on this highest architecture (with 16 cores and 1000 columns) using the proposed *HSAP* algorithm. The energy reference is the energy budget E^{max}.

5.4.5.1 Single Application Results

For all single applications considered, the *ECosynth* results in the least energy consumption (for energy consumption, lower is better). This is expected as the primary objective of this technique is to minimize energy. However, the lifetime (measured as MTTF) obtained using this technique is the least (for MTTF, higher is better). For the *MPEG4 decoder* application, the *ECosynth* results in 44% lower MTTF (row 3, column 6) and 25% lower energy (row 3, column 5) as compared to the reference approach. The *RPGen* technique determines the number of cores based on reliability with energy as a constraint. This technique does not consider reconfigurable area, and therefore the entries reporting the reconfigurable area usage (row 4, columns 4) is 0. The lifetime reliability using this technique is on average better than *ECosynth* by 30% as this is explicitly maximized. The energy consumption of this technique is, however, higher by 33%. This high energy overhead of *RPGen* is attributed to the fact that the underlying architecture for this technique is the static multiprocessor system with no reconfigurable area. Therefore, the technique does not benefit from the lower energy consumption of the hardware

Table 5.1 Platform determination with area, energy, and reliability results

Single applications

Applications	Technique	Cores	RA size	Energy	MTTF
MPEG4 decoder	ECosynth	9	200	0.75	0.56
	RPGen	16	0	1.00	0.73
	Proposed	8	300	0.90	0.94
JPEG decoder	ECosynth	2	100	0.79	0.68
	RPGen	4	0	0.96	0.81
	Proposed	4	200	0.91	1.00
Final platform	ECosynth	9	600	–	–
	RPGen	16	0	–	–
	Proposed	8	500	–	–

Use-cases

Applications	Technique	Cores	RA size	Energy	MTTF
usecase_1	ECosynth	16	700	0.81	0.51
	RPGen	16	0	1.00	0.58
	Proposed	12	600	1.00	0.90
usecase_2	ECosynth	10	1000	0.85	0.60
	RPGen	16	0	1.00	0.68
	Proposed	10	800	0.90	1.00
Final platform	ECosynth	16	1000	–	–
	RPGen	16	0	–	–
	Proposed	12	800	–	–

implementation of actors. In comparison to both these techniques, the proposed technique results in the highest MTTF with an improvement of 28% with respect to *RPGen* and 67% with respect to *ECosynth*. In terms of energy consumption, the proposed technique consumes 10% lower energy than *RPGen*. Finally, the resource utilization of the proposed technique is also better than the existing techniques. A similar trend is observed for all the single applications considered (including those not shown in this table). On average for all these applications, the proposed technique improves lifetime by an average 30% with respect to *RPGen* and 65% with respect to *ECosynth*.

The minimum resources required for all the single applications are reported in the table at rows 9–11, columns 3–4. As seen from these entries, the proposed approach satisfies the resource constraint and also leads to the minimum resource usage.

5.4.5.2   Use-Case Results

To demonstrate the performance of the ILP-based Pareto merging technique for use-cases, an experiment is conducted with ten synthetic use-cases. Two of these are reported in the table. The compositions of these use-cases are as follows: $usecase\_1 = \langle JPEGDec, MP3Dec \rangle$ and $usecase\_2 = \langle MPEG4Dec, SRC, H263Dec \rangle$. Since none of the two existing works consider use-cases for platform determination, the resource usage for these techniques is performed by optimizing the concurrent applications individually, with the resource constraint modified according to their throughput requirements. An example is illustrated below.

Let $usecase\_i = \langle App\_A, App\_B \rangle$ with throughput constraint of App_A be 1.5 times the throughput constraint of App_B. The overall resource constraint of 16 cores with 1000 reconfigurable columns is distributed to these applications such that the constraint on App_A is 9 cores with 600 columns, and that of application App_B is 7 cores with 400 columns. The proposed approach uses ILP based Pareto merging to determine the resource requirement for every use-case. As seen from the table, for $usecase\_1$, the *ECosynth* still achieves the minimum energy consumption. The MTTF is still the least. The *RPGen* although maximizes MTTF, the improvement over that of *ECosynth* is only 13%. This is due to the non-availability of the reconfigurable area leading to the missing reliability and reconfigurable area trade-off. The proposed approach improves lifetime by 55% as compared to *RPGen*. A similar trend is observed for the second use-case. On average for all the ten use-cases considered, the proposed approach improves lifetime by 50% with respect to *RPGen* and 70% with respect to *ECosynth*, while satisfying the given area, energy, and performance budget.

The maximum resource used for all the ten use-cases (including the two shown in the table) is reported in rows 9–11, columns 9–10. As seen from these results, the proposed approach minimizes the number of processing cores by 25% and reconfigurable area usage by 20% with respect to *ECosynth*, while maximizing the lifetime by an average 65% for single application and an average 70% for use-cases and satisfying the area, energy, and performance constraint. With respect to the lifetime maximum technique (*RPGen*), the proposed co-design approach improves lifetime by an average 30% for single applications and an average 50% for use-cases.

5.5   Remarks

This work presented a fast heuristic for hardware–software task partitioning for reconfigurable multiprocessor systems. The objective is to improve the transient fault-tolerance of the system together with the lifetime reliability of the cores. Based on this heuristic, a hardware–software co-design technique is proposed that determines the minimum resources needed to maximize the reliability while satisfying the given energy, cost, and performance constraint. The co-design methodology incorporates integer linear programming to merge the Pareto-points of the individual applications to determine the resource usage for concurrent applications (use-cases). Experiments conducted with synthetic and real-life application SDFGs demonstrate that the proposed hardware–software partitioning technique maximizes lifetime reliability by 10% for a stringent transient fault-tolerance requirement. With relaxed requirements, the proposed approach is able to improve lifetime by an average 60%. Moreover, the proposed hardware–software co-design approach improves the lifetime reliability by an average 70%, while consuming 25% fewer cores and 20% lower reconfigurable area size as compared to the existing technique.

References

1. P. Axer, M. Sebastian, R. Ernst, Reliability analysis for MPSoCs with mixed-critical, hard real-time constraints, in *Proceedings of the Conference on Hardware/Software Codesign and System Synthesis (CODES+ISSS)* (ACM, 2011), pp. 149–158
2. D. Bertozzi, A. Jalabert, S. Murali, R. Tamhankar, S. Stergiou, L. Benini, G. De Micheli, NoC synthesis flow for customized domain specific multiprocessor systems-on-chip. IEEE Trans. Parallel Distrib. Syst. (TPDS) **16**(2), 113–129 (2005)
3. A. Das, A. Kumar, B. Veeravalli, Aging-aware hardware-software task partitioning for reliable reconfigurable multiprocessor systems, in *Proceedings of the International Conference on Compilers, Architecturesand Synthesis for Embedded Systems (CASES)* (IEEE Press, 2013), pp. 1:1–1:10
4. A. Das, A. Kumar, B. Veeravalli, Energy-aware task mapping and scheduling for reliable embedded computing systems. ACM Trans. Embed. Comput. Syst. (TECS) **13**(2s), 72:1–72:27 (2014)
5. J. Huang, J.O. Blech, A. Raabe, C. Buckl, A. Knoll, Analysis and optimization of fault-tolerant task scheduling on multiprocessor embedded systems, in *Proceedings of the Conference on Hardware/Software Codesign and System Synthesis (CODES+ISSS)* (ACM, 2011), pp. 247–256

6. L. Huang, F. Yuan, Q. Xu, On task allocation and scheduling for lifetime extension of platform-based MPSoC designs. IEEE Trans. Parallel Distrib. Syst. (TPDS) **22**(12), pp. 2088–2099 (2011)

7. L. Jiashu, A. Das, A. Kumar, A design flow for partially reconfigurable heterogeneous multi-processor platforms, in *Proceedings of the International Symposium on Rapid System Prototyping (RSP)* (IEEE, 2012) pp. 170–176

8. C. Krishna, A. Singh, Reliability of checkpointed real-time systems using time redundancy. IEEE Trans. Reliab. **42**(3), 427–435 (1993)

9. S.-W. Kwak, B.-J. Choi, B.-K. Kim, An optimal checkpointing-strategy for real-time control systems under transient faults. IEEE Trans. Reliab. **50**(3), 293–301 (2001)

10. P.K. Saraswat, P. Pop, J. Madsen, Task mapping and bandwidth reservation for mixed hard/soft fault-tolerant embedded systems, in *Proceedings of the IEEE Symposium on Real-Time and Embedded Technology and Applications (RTAS)* (IEEE Computer Society, 2010), pp. 89–98

11. M.T. Schmitz, B.M. Al-Hashimi, P. Eles, Cosynthesis of energy-efficient multimode embedded systems with consideration of mode-execution probabilities. IEEE Trans. Comput. Aided Des. Integr. Circuits Syst. (TCAD) **24**(2), 153–169 (2005)

12. S. Stuijk, M. Geilen, T. Basten, SDF3: SDF for free, in *Proceedings of the International Conference on Application of Concurrency to System Design (ACSD)* (IEEE Computer Society, 2006), pp. 276–278

13. J. Teich, Hardware/software codesign: the past, the present, and predicting the future. Proc. IEEE **100**(Special Centennial Issue), 1411–1430 (2012)

14. C. Zhu, Z.P. Gu, R.P. Dick, L. Shang, Reliable multiprocessor system-on-chip synthesis, in *Proceedings of the Conference on Hardware/Software Codesign and System Synthesis (CODES+ISSS)* (ACM, 2007), pp. 239–244

Chapter 6
Design-Time Analysis for Fault-Tolerance

6.1 Introduction

This chapter solves the following problem. Given a heterogeneous multiprocessor system and a set of multimedia applications, how to assign and order tasks of every application on the component cores such that the total energy consumption is minimized while guaranteeing to satisfy performance requirements of these application under all possible fault-scenarios. The scope of this work is limited to permanent failures of processing cores. Following are the key contributions of this chapter:

- task mapping technique to minimize energy consumption for every fault-scenarios, satisfying performance requirement;
- a scheduling technique to minimize run-time schedule construction and storage overhead; and
- a heuristic to minimize design space exploration time to find an energy-efficient mapping.

The remainder of this chapter is organized as follows. The problem formulation is discussed in Sect. 6.2, followed by the proposed design methodology in Sect. 6.3. The energy minimum task mapping technique for different fault-scenarios is discussed next in Sect. 6.4 and the proposed self-timed execution-based scheduling technique in Sect. 6.5. Experimental setup and results are discussed in Sect. 6.6. Lastly, conclusions are presented in Sect. 6.7.

© Springer International Publishing AG 2018
A.K. Das et al., *Reliable and Energy Efficient Streaming Multiprocessor Systems*, Embedded Systems, https://doi.org/10.1007/978-3-319-69374-3_6

6.2   Problem Formulation

6.2.1   Application and Architecture Model

An application is represented as synchronous data flow graphs (SDFGs) $\mathscr{G}_{\mathrm{app}} = (\mathbb{A}, \mathscr{C})$ consisting of a finite set $\mathbb{A}$ of actors and a finite set $\mathscr{C}$ of channels. Every actor $a_i \in \mathbb{A}$ is a tuple (t_i, μ_i), where t_i is the execution time of a_i and μ_i is its state space (program and data memory). Number of actors in an SDFG is denoted by N_a where $N_a = |\mathbb{A}|$. Performance of an SDFG is specified in terms of throughput constraint $\mathbb{T}_c$. The architecture is represented as a graph $\mathscr{G}_{\mathrm{arc}} = (\mathbb{C}, \mathbb{E})$, where $\mathbb{C}$ is the set of nodes representing cores of the architecture and $\mathbb{E}$ is the set of edges representing communication channels among the cores. The total number of cores is denoted by N_c, i.e., $N_c = |\mathbb{C}|$. Each core $c_j \in \mathbb{C}$ is a tuple $\langle h_j, O_j \rangle$, where h_j represents the heterogeneity type of c_j, and O_j is the set of voltages and frequency pairs supported on c_j.

6.2.2   Mapping Representation

The following notations are defined.

N_o	maximum number of operating points of a core
M_n	mapping of $\mathscr{G}_{\mathrm{app}}$ on $\mathscr{G}_{\mathrm{arc}}$ with n cores where $n \leq N_c$
ϕ_i	core on which actor a_i is mapped in mapping M_n
θ_i	frequency assigned to actor a_i
Ψ_j	set of actors mapped to core c_j
s_f	fault-scenario with f faulty cores $= \langle c_{i_1}, c_{i_2}, \ldots, c_{i_f} \rangle$

The objective of the optimization problem is to minimize energy consumption for all fault-scenarios by solving the following:

- *actor distribution:* i.e., to determine the assignment of the actors of the SDFG on the cores of the multiprocessor system;
- *operating point:* i.e., to determine the voltage and frequency of the cores for executing the actors of the SDFG.

The mapping is represented as $M_n = \left(\mathscr{M}_d^n \ \mathscr{M}_o \right)$. Two variables $x_{i,j}$ (representing the *actor distribution*) and $y_{i,k}$ (representing the *operating point*) are defined as follows.

$$x_{i,j} = \begin{cases} 1 & \text{if actor } \boldsymbol{a}_i \text{ is executed on core } \boldsymbol{c}_j \\ 0 & \text{otherwise} \end{cases} \qquad y_{i,k} = \begin{cases} 1 & \text{if actor } \boldsymbol{a}_i \text{ is executed at} \\ & \text{operating point } o_k \\ 0 & \text{otherwise} \end{cases}$$

Constraints on these variables are set such that an actor is mapped to only one core at a single operating point. Thus,

$$\sum_{j=0}^{N_c-1} x_{i,j} = 1 \text{ and } \sum_{k=0}^{N_o-1} y_{i,k} = 1 \quad \forall \boldsymbol{a}_i \in \mathbb{A} \tag{6.1}$$

The *actor distribution* and *operating point* of an SDFG are represented as two matrices:

$$\mathcal{M}_d^n = \begin{pmatrix} x_{0,0} & x_{0,1} & \cdots & x_{0,n-1} \\ x_{1,0} & x_{1,1} & \cdots & x_{1,n-1} \\ \vdots & \vdots & \ddots & \vdots \\ x_{N_a-1,0} & x_{N_a-1,1} & \cdots & x_{N_a-1,n-1} \end{pmatrix} \text{ and } \mathcal{M}_o = \begin{pmatrix} y_{0,0} & y_{0,1} & \cdots & y_{0,N_o-1} \\ y_{1,0} & y_{1,1} & \cdots & y_{1,N_o-1} \\ \vdots & \vdots & \ddots & \vdots \\ y_{N_a-1,0} & y_{N_a-1,1} & \cdots & y_{N_a-1,N_o-1} \end{pmatrix}$$

$$\tag{6.2}$$

The core assignment and operating point of actor $\boldsymbol{a}_i$ are given by

$$\phi_i = \mathbf{X}_i \times \mathbb{N}_{N_c} \text{ where } \mathbf{X}_i = \begin{pmatrix} x_{i,0} & x_{i,1} & \cdots & x_{i,N_c-1} \end{pmatrix} \text{ and } \mathbb{N}_{N_c} = \begin{pmatrix} 0 & 1 & \cdots & N_c-1 \end{pmatrix}^T$$

$$\theta_i = \mathbf{Y}_i \times \mathbb{N}_{N_o} \text{ where } \mathbf{Y}_i = \begin{pmatrix} y_{i,0} & y_{i,1} & \cdots & y_{i,N_o-1} \end{pmatrix} \text{ and } \mathbb{N}_{N_o} = \begin{pmatrix} 0 & 1 & \cdots & N_o-1 \end{pmatrix}^T$$

6.2.3 Mapping Encoding

A unique ID is assigned to each mapping M_n as calculated in Eq. (6.3).

$$mID(M_n) = \sum_{i=1}^{N_a} \phi_i \cdot (N_c)^i \tag{6.3}$$

6.2.4 Energy Modeling

The computation and the communication energy are reused from Chap. 4; this section provides the modeling of energy associated with task-migration. Migration overhead is defined as the overhead associated with moving from one mapping to another. This is governed by two quantities—the state space of the actors(s) participating in the migration process and the distance (number of hops) through which the state space is migrated. The state space of an actor consists of the data memory and the pre-compiled object code. It is assumed that the task is compiled for all the h different heterogeneity types. As an example, if the platform consists of both RISC and CISC architectures, the code for a task needs to be pre-compiled for these two ISAs and stored in the memory. The state space then consists of compiled code for both these heterogeneity.

It is assumed that a given multiprocessor system consists of one or more task migration modules (TMMs), which can access the memory of a core without interfering with its operation. For these systems, the state space of an actor (on a faulty core) can be recovered and hence migrated to some other core. For multiprocessor systems without TMMs, task migration involves migrating the state space of an actor from a shared memory to the new core where it is to be mapped. The migration overhead is represented as energy and is termed migration energy.

$$MigrationEnergy = \sum_{\forall a_i \in \Psi_j} S_i \cdot E_{\text{bit}}(\phi_i^{\text{init}}, \phi_i^{\text{final}}) \qquad (6.4)$$

where S_i is the state space of actor a_i, ϕ_i^{init} and ϕ_i^{final} are the cores on which actor a_i is mapped before and after task migration, respectively and E_{bit} is the energy required to communicate every bit of data across the NoC (Eq. (4.9)).

6.3   Design Methodology

The fault-tolerant task mapping methodology consists of two phases—analysis of applications at design-time and execution at run-time. The focus of this work is on the design-time analysis; however, for the sake of completeness, a brief overview is provided on how to use the design-time analysis results at run-time.

The fault-tolerant task mapping methodology is outlined in Fig. 6.1. For every fault-scenario with f faulty cores, an optimal mapping is generated that satisfies the throughput requirement and results in minimum energy consumption. These mappings are encoded by the *Encode Mapping* block to generate a unique mapping ID and is stored in memory corresponding to the fault-scenario. An example snippet of this memory is shown at the bottom right corner of the figure. In this snippet, fault-scenario 0 (meaning core 0 is faulty) has mapping ID 189, while fault-scenario 0-1 has mapping ID 39. At run-time, an application is executed until a fault is detected. On detection,[1] the corresponding fault-scenario is identified and the encoded mapping is fetched from the memory. This mapping is then decoded by the *Decode Mapping* block and forwarded to the *Task Migration* module where the actual task migration is performed.

---

[1]This work is orthogonal to any fault-detection mechanism.

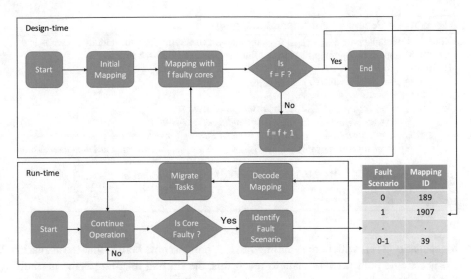

Fig. 6.1 Proposed design methodology

6.4   Fault-Tolerant Mapping Generation

Fault-tolerant mappings are generated using Algorithm 9. There are F stages of the algorithm, where F is the maximum number of faults that can be tolerated on the platform. At every stage f ($1 \leq f \leq F$), mappings are generated—one for each fault-scenario with f faulty cores.

Each stage of the algorithm has multiple sub-steps. The first sub-step is the generation of a set (S^f) of fault-scenarios (line 2). There are $^{N_c}P_f$ fault-scenarios in this set. An example set with 2 out of 3 cores as faulty ($f = 2, N_c = 3$) is the set $S^f = \{\langle 0, 1 \rangle, \langle 1, 0 \rangle, \langle 0, 2 \rangle, \langle 2, 0 \rangle, \langle 1, 2 \rangle, \langle 2, 1 \rangle\}$.[2] For every fault-scenario of the set S^f, the last core (c_{i_f}) of the tuple $\langle c_{i_1}, c_{i_2}, \ldots, c_{i_f} \rangle$ is considered as the current faulty core, and a lower order tuple is generated by omitting c_{i_f} (line 5). This gives fault-scenario s_{f-1} with $f - 1$ faulty cores for which the optimal mapping is already computed (and stored in *HashMap*) in the previous stage (i.e., at stage $f - 1$). This assumption is based on the fact that at most one core can fail at any given time. An example is provided to better clarify this. The fault-scenario $\langle 3, 1, 5 \rangle$ implies that faults occurred first on core c_3, followed by on core c_1, and finally on core c_5. To reach this fault-scenario, the system need to encounter fault-scenario $\langle 3, 1 \rangle$ first. Mapping for $\langle 3, 1 \rangle$ is therefore considered as the starting mapping for $\langle 3, 1, 5 \rangle$ with core c_5 as current failing core. Similarly, mapping for $\langle 3 \rangle$ is the starting mapping

---

[2]A fault-scenario (0,1) implies fault occurring first at core c_0 and then at core c_1. Thus, fault-scenario (0,1) is different from fault-scenario (1,0) implying a permutation in the fault-scenario computation.

Algorithm 9 Generate fault-tolerant mappings

Require: Initial mapping $M_{n_{arc}}$, G_{app}, G_{arc}, throughput constraint $\mathbb{T}_c$, fault-tolerance level F
Ensure: Minimum energy mappings for all fault-scenarios with $f = 1$ to F faults
 1: **for** $f = 1$ to F **do**
 2:    $S^f = genFaultScenarios(f)$
 3:    **for** $s_f \in S^f$ **do**
 4:       $s_f = (c_{i_1}, c_{i_2}, \cdots, c_{i_{f-1}}, c_{i_f})$ //represent fault-scenario
 5:       $s_{f-1} = (c_{i_1}, c_{i_2}, \cdots, c_{i_{f-1}})$ //generate reduced fault-scenario
 6:       $M_{f-1} = HashMap[s_{f-1}].getMap()$ //fetch mapping for reduced fault-scenario
 7:       $M_f = genMinEnergyMap(M_{f-1}, G_{app}, G_{arc}, \mathbb{T}_c, c_{i_f}, s_f)$ //generate minimum energy map

 8:       $HashMap[s_f].setMap(M_f)$ //store mapping for the fault-scenario
 9:    **end for**
10: **end for**

---

for scenario $\langle 3, 1 \rangle$, with core c_1 failing next. A point to note here is that the scenario $\langle 3 \rangle$ is a single fault-scenario, and to reach this, the starting mapping is the no fault initial mapping M_{N_c}.

An important component of Algorithm 9 is the generation of minimum energy mapping *genMinEnergyMap()*. This routine takes a starting mapping M_n, the current faulty core (c_j), and the fault-scenario (s_f) and generates a new mapping M_{n-1} with core c_j as faulty. This new mapping satisfies throughput constraint and gives minimum energy (computation and communication). Details of this routine are provided in the next section. Once an optimal mapping is determined (line 7), the algorithm stores it in the *HashMap* for the particular fault-scenario (line 8). This is repeated for every scenario of set S^f.

6.4.1 Minimum Energy Mapping Generation

A heuristic is proposed to determine mapping and scheduling of applications on a multiprocessor system. This is shown as a pseudo-code in Algorithm 10. The algorithm has two parts—remapping mandatory actors (line 1) and searching for the minimum energy mapping (lines 4–16). Mandatory remapping is generated by moving the actors on the faulty core only (c_ϑ). These actors are part of the set $\Psi(c_\vartheta)$. Actor remapping is accomplished by selecting $|\Psi(c_\vartheta)|$ cores (not all different) from the set of operating cores $\mathbb{C} \setminus s_f$ to remap all $a_i \in \Psi(c_\vartheta)$. The number of possible mappings can be calculated by considering the equivalent problem of determining the number of ways of choosing a sample of $|\Psi(c_\vartheta)|$ balls with replacement from a set of $|\mathbb{C} \setminus s_f|$ balls. This is equal to $|\mathbb{C} \setminus s_f|^{|\Psi(c_\vartheta)|}$. The generated mappings are pruned according to standard speed-up techniques (such as processor load [8]). These mappings are stored in an array Γ_ϑ, and the array is sorted in terms of communication energy (line 2). The *maxIter* best mappings are selected and used in the next stage. This number (*maxIter*) determines termination of the algorithm.

Algorithm 10 *GenMinEnergyMap*(): energy aware mapping

Require: Mapping M_n, $\mathcal{G}_{app}$, $\mathcal{G}_{arc}$, throughput constraint $\mathbb{T}_c$, faulty core c_ϑ and fault-scenario s_f
Ensure: New mapping M_{n-1}
1: Γ_ϑ = Set of mappings generated from M_n by remapping all $a_i \in \Psi(c_\vartheta)$ to some $c_j \in \mathbb{C} \setminus s_f$
2: Sort the mappings in Γ_ϑ according to communication energy and $M_{temp} = \Gamma_\vartheta[0]$
3: Initialize *numIter* = 0; $M_{best} = M_{temp}$; $E_{best} = calcEnergy(M_{temp})$
4: **while** *numIter* $\leq$ *maxIter* **do**
5:      $[i \, j \, k]$ = *RemapActor*(M_{temp}, $\mathcal{G}_{app}$, $\mathcal{G}_{arc}$, $\mathbb{T}_c$, s_f)
6:      **if** $i \geq 0$ **then**
7:          $x_{ij} = 1$ and $x_{ij'} = 0 \; \forall j' \neq j$; $y_{ik} = 1$ and $y_{ik'} = 0 \; \forall k' \neq k$; Update M_{temp}
8:      **else**
9:          $E = calcEnergy(M_{temp})$
10:         **if** $E < E_{best}$ **then**
11:             $M_{best} = M_{temp}$; $E_{best} = E$
12:         **end if**
13:         *numIter* + +
14:         $M_{temp} = \Gamma_\vartheta[numIter]$
15:     **end if**
16: **end while**
17: Return M_{best}

The minimum energy search part of the algorithm (lines 4–16) remaps one or more actors selectively to determine the minimum energy. At each iteration, the starting mapping is one of the mappings of set Γ_ϑ. The *RemapActor*() routine selects an actor to be remapped, satisfying throughput requirement. If the return set is nonempty (implying actors can be remapped without violating the throughput constraint), the actor is remapped to a core at an *operating point* determined by the *RemapActor*() routine (line 7). The process is continued as long as no actors can be remapped without violating the throughput. The total energy of the mapping is calculated using the *calcEnergy*() routine that incorporates (1) the computation energy; (2) the communication energy; and (3) the migration energy.

If the total energy of the mapping is lower than the minimum energy (E_{best}) obtained thus far in the algorithm, the best values are updated (line 11). The number of iterations is incremented (line 13) and the whole search is repeated starting from the next mapping in the set Γ_ϑ. The algorithm terminates when *numIter* becomes equal to *maxIter* and the best mapping is returned (line 17).

6.4.2   Minimum Energy Actor Remapping

Algorithm 11 provides the pseudo-code for the *RemapActor*() subroutine that uses a gradient function to evaluate each actor assignment. The total energy and throughput are evaluated by assigning every actor to every core at every operating point (line 6). The MSDF[3] tool is used to compute the schedule and throughput from a given mapping. If the throughput for the new assignment is greater than the throughput

---

Algorithm 11 *RemapActor*(): remap actors to minimize energy

Require: Mapping M, $\mathcal{G}_{app}$, $\mathcal{G}_{arc}$, throughput constraint $\mathbb{T}_c$, fault-scenario s_f
Ensure: Determine an actor to be remapped, the corresponding core and *operating point*
1: $E := calcEnergy(M)$; $[\mathcal{S}\ \mathbb{T}] = MSDF^3(M, \mathcal{G}_{app}, \mathcal{G}_{arc}, \mathbb{T}_c)$; $G_{best} = 0$; $i_{best} = j_{best} = k_{best} = -1$
2: **for all** $a_i \in \mathbb{A}$ **do**
3:    **for all** $c_j \in \mathbb{C} \setminus s_f$ **do**
4:       **for all** $k \in [0, 1, \cdots N_f - 1)$ **do**
5:          $M_{new} = M$; Set $x_{ij} = y_{ik} = 1$ and $x_{ij'} = y_{ik'} = 0\ \forall j' \neq j$ and $k' \neq k$
6:          $[\mathcal{S}_{new}\ \mathbb{T}_{new}] = MSDF^3(M_{new}, \mathcal{G}_{app}, \mathcal{G}_{arc}, \mathbb{T}_c)$; $E = calcEnergy(M_{new})$
7:          **if** $((\mathbb{T}_{new} \geq \mathbb{T}_c)$ && $(E_{new} < E))$ **then**
8:            $G = \frac{E - E_{new}}{\mathbb{T}_{new} - \mathbb{T}}$
9:            **if** $G > G_{best}$ **then**
10:              $G_{best} = G$; $i_{best} = i$; $j_{best} = j$; $k_{best} = k$
11:            **end if**
12:          **end if**
13:       **end for**
14:    **end for**
15: **end for**
16: return $[i_{best}\ j_{best}\ k_{best}]$

Algorithm 12 Generate initial mapping

Require: $\mathcal{G}_{app}$, $\mathcal{G}_{arc}$ and throughput constraint $\mathbb{T}_c$
Ensure: Minimum energy initial mapping M
1: Initialize $[M\ \mathcal{S}\ \mathbb{T}] = SDF^3(\mathcal{G}_{app}, \mathcal{G}_{arc}, \mathbb{T}_c)$
2: **while** true **do**
3:    $[i\ j\ k] = RemapActor(M, \mathcal{G}_{app}, \mathcal{G}_{arc}, \mathbb{T}_c, \emptyset)$
4:    **if** $i \geq 0$ **then**
5:       Update M with $x_{ij} = y_{ik} = 1$ and $x_{ij'} = y_{ik'} = 0\ \forall j' \neq j$ and $k' \neq k$
6:    **else**
7:       break
8:    **end if**
9: **end while**
10: Return M

---

constraint and the energy is lower than the energy of the initial mapping M, the gradient is computed (line 8). If the gradient is higher than the best gradient obtained thus far, the best values are updated (line 11). The best actor, core, and operating point are returned as output of the algorithm.

6.4.3 Initial Mapping Generation

Algorithm 12 provides the pseudo-code for the initial mapping generation procedure of our proposed methodology. The initial mapping (at line 1) is obtained by the SDF^3 tool for SDFGs. The *RemapActor*() routine selects one actor to be remapped to a core at a frequency such that energy is minimized with least throughput degradation. This is similar to Algorithm 10 with the all working cores ($s_f = \emptyset$).

6.5   Fault-Tolerant Scheduling

None of the existing fault-tolerant techniques address scheduling. If the run-time schedule is different from that used for analysis at design-time, the throughput obtained will be significantly different than what is guaranteed at design-time. There are two solutions to the scheduling problem:

- store actor mapping and scheduling for all fault-scenarios and for all applications from design-time (henceforth referred to as *storage-based* solution); and
- construct schedule at run-time based on the mappings stored from the design-time (henceforth referred to as *construction-based* solution).

The former is associated with high storage overhead and the latter with longer execution time. Both storage and execution time overhead are crucial for streaming applications. Here we propose a self-timed execution based scheduling to solve the two problems. Based on the basic properties of self-timed scheduling, it can be proven if schedule of actors on a uniprocessor system is used to derive the schedules for a multiprocessor system maintaining actor firing order, the resultant multiprocessor schedule would be deadlock-free [2]. However, the throughput obtained using this technique could be lower than the maximum throughput of a multiprocessor schedule constructed independently. Thus, as long as this throughput deviation is bounded, the schedule for any processor could be easily constructed from the mapping of actors to this processor and a given uniprocessor schedule.

Figure 6.2 shows how to construct the schedules using our proposed scheduling technique. The actor-core mapping indicates actors a_0, a_1, and a_3 are mapped to core 0. The initial steady-state schedule indicates that there are two instances of a_1 and one each for actors a_0 and a_3, respectively. The steady-state order of actor firing on core 0 is determined from this initial schedule by retaining only the mapped actors. In a similar way, the steady-state schedules are constructed for all other processors. The transient part of the schedules are constructed from a given initial uniprocessor transient schedule by retaining the mapped actors. However, the only difference of the transient-phase schedule construction with the steady-state phase is that for the transient phase, the number of actors firing is important and not the exact order. This is indicated by a number against each actor for each processor, as seen in the figure.

During the steady-state operation, every core maintains counts of the number of remaining steady-state firings for the actors mapped to the core. These numbers are updated when an actor completes its execution. When a fault occurs, the mapped actors on the faulty core are moved to new location(s) (cores), along with the remaining firing count. On such cores, which have at least one incoming migrated actor, all actors are allowed to execute in a self-timed manner to finish the remaining firing counts of the current pending iteration (as in the initial transient phase). Subsequently, steady-state order can be enforced for the moved actors. This will prevent the application from going into deadlock when a fault occurs. In determining actor counting in the steady-state iterations, schedule minimization is disabled. As an example, in Fig. 6.2, the steady state schedule constructed for core 2 consists of two executions of actor a_5 as opposed to one in the otherwise minimized schedule.

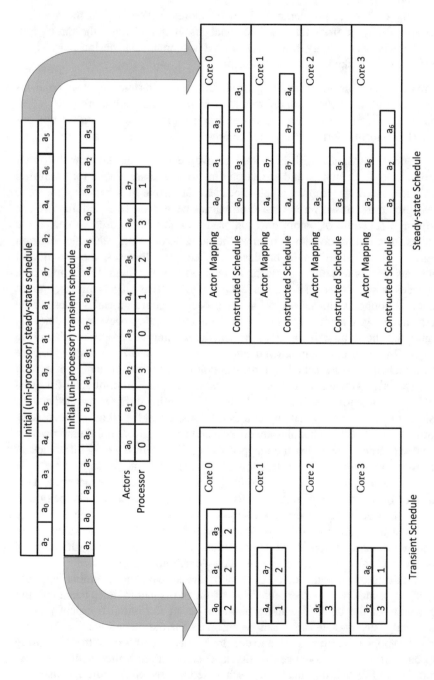

Fig. 6.2 Constructing transient and steady-state schedules from initial schedule and actor allocation [5]

Algorithm 13 Schedule generation

Require: $\mathcal{G}_{app}, \mathcal{G}_{arc}, \mathbb{T}_c, N_s$ and Δ
Ensure: Schedule for all fault-scenarios
1: **forall** $f \in [1 \cdots F]$ **do** $S^f = genFaultScenarios(f)$
2: $maxIter = |S^f|$; $sDB = constructUniSchedule(\mathcal{G}_{app}, N_s)$; $sDB_t = sDB$
3: **while** $S^f \neq \emptyset$ **do**
4:   **for all** schedule $l_i \in sDB$ **do**
5:     Initialize $count = 0$
6:     **for all** $s_f \in S^f$ **do**
7:       $M_{temp} = HashMap[s_f].getMap()$; $\mathbb{T} = SMSDF^3(M_{temp}, l_i, \mathcal{G}_{app}, \mathcal{G}_{arc})$
8:       **if** $\mathbb{T} \geq \mathbb{T}_c$ **then**
9:         $count ++$
10:      **end if**
11:    **end for**
12:    $l_i.rank = count$
13: **end for**
14: $l_{min} = getHighestRankSchedule(sDB)$
15: **for all** $s_f \in S^f$ **do**
16:    $M_{temp} = HashMap[s_f].getMap()$; $\mathbb{T} = SMSDF^3(M_{temp}, l_{min}, \mathcal{G}_{app}, \mathcal{G}_{arc})$
17:    **if** $\mathbb{T} \geq \mathbb{T}_c$ **then**
18:      $Sche[s_f] = l_{min}$; $S^f.eliminate(s_f)$
19:    **end if**
20: **end for**
21: $numIter ++$
22: **if** $numIter > maxIter$ **then**
23:    $numIter = 0$; $C = C - \Delta$
24: **end if**
25: **end while**

Algorithm 13 provides the pseudo-code for the modified self-timed execution technique for generating steady-state schedules. The first step is the construction of uni-processor schedules (line 2). A list scheduling technique is used for this purpose along with several algorithms for tie-breaking, for example, ETF (earliest task first), DLS (dynamic level scheduling), etc. These algorithms are implemented in the *constructUniSchedule()* routine. The number of uni-processor schedules constructed using this routine is a user-defined parameter N_s. These schedules are stored in a database in memory (*sDB*). The list of fault-scenarios possible with F faults are also listed in the set S^f. Using each uni-processor schedule as the initial schedule, throughput is computed for the given application for all fault-scenario mappings. The SMSDF[3] computes the throughput of a mapping using a given uni-processor schedule.

For each uni-processor schedule from *sDB*, a rank is calculated (lines 4–13). Rank of a schedule is the number of fault-scenarios that can be addressed using this as the initial schedule. A schedule with the highest rank is selected and assigned as the initial schedule for the successful fault-scenarios (lines 14–20). The successful fault-scenarios are discarded from the list of fault-scenarios (S^f). The process is repeated as long as the set S^f is non-empty. The limited set of uni-processor schedules does not guarantee throughput satisfiability for all fault-scenarios. If such a fault-scenario exists, S^f is never Ø causing the algorithm to be stuck in a loop. To avoid such situations, a check is performed (lines 22–24) to limit the number of iterations. The maximum number of iterations is upper bounded by the number of fault-scenarios. Every time the iteration count reaches this value, the throughput constraint is decremented by a small quantity, Δ. The algorithm thus allows for a graceful performance degradation. The granularity of this is based on the execution time and solution quality trade-off.

6.6 Results

Experiments are conducted with synthetic and real application SDFGs on Intel Xeon 2.4 GHz server running Ubuntu Linux. Fifty synthetic applications are generated with the number of actors in each application selected randomly from the range 8–25. Additionally, fifteen real applications are considered with seven from streaming and the remaining eight from non-streaming domain. The streaming applications are obtained from the benchmarks provided in the SDF^3 tool [10]. These are *H.263 Encoder*, *H.263 Decoder*, *H.264 Encoder*, *MP3 Decoder*, *MPEG4 Decoder*, *JPEG Decoder*, and *Sample Rate Converter*. The non-streaming application graphs considered are *FFT*, *Romberg Integration*, and *VOPD* from [1] and one application each from *automotive*, *consumer*, *networking*, *telecom*, and *office automation* benchmark suite [6]. These applications are executed on a multiprocessor system with cores arranged in a mesh-based topology. A heterogeneity of 3 ($h = 3$) is assumed for the cores, i.e., each core can be one of three different types. Four *operating points* are assumed for each core.

We compare our approach with (1) the throughput maximization technique of [9] (referred to as *TMax*), (2) the migration overhead minimization technique of [11] (referred to as *OMin*), (3) the energy minimization technique of [7] (referred to as *EMin*), (4) the throughput constrained migration overhead minimization technique of [4] (referred to as *TConOMin*), and (5) the throughput constrained communication energy minimization technique of [3] (referred to as *TConCMin*). The technique proposed here minimizes total energy (computation and communication energy) with throughput as a constraint and is referred to as *TConEMin*.

6.6.1 Algorithmic Complexity Analysis

There are three algorithms proposed in this work—fault-tolerant mapping generation algorithm (Algorithms 9–11), the initial mapping generation algorithm (Algorithm 12), and the schedule generation algorithm (Algorithm 13). Complexity of Algorithm 9 is calculated as follows. The number of iterations of the algorithm is determined by the number of fault-scenarios with F faults, given by

$$N_{\text{FS}} = \sum_{f=1}^{F} {}^{N_c}P_f \tag{6.5}$$

At each iteration, the *genMinEnergyMap* algorithm is invoked. Complexity of Algorithm 9 is given by Eq. (6.6) where C_{10} is the complexity of Algorithm 10.

$$O(C_9) = O\left(N_{\text{FS}} \cdot O(genMinEnergyMap)\right) = O(N_{\text{FS}} \cdot C_{10}) \tag{6.6}$$

Complexity of Algorithm 10 is governed by two factors—parameter *maxIter* and the routine *RemapActor()*. Core and frequency assignments for an actors are accomplished in constant time. Assuming the *RemapActor()* routine to be executed η times on average for each value of *numIter*, the complexity of Algorithm 10 is

$$C_{10} = maxIter \cdot \eta \cdot O(RemapActor) \tag{6.7}$$

The *RemapActor()* routine remaps each actor on each functional core at each frequency to determine if the throughput constraint is satisfied and the energy is lower than the minimum energy obtained thus far. If actor assignment operations take unit time and the complexity of the MSDF^3 engine is denoted by $O(\text{MSDF}^3)$, the overall complexity of Algorithm 11 is given by Eq. (6.8).

$$O(RemapTask) = C_{11} = O(N_a \cdot N_c \cdot N_o \cdot O(\text{MSDF}^3)) \tag{6.8}$$

Combining Eqs. (6.6)–(6.8), the complexity of the fault-tolerant mapping generation algorithm is given by Eq. (6.9).

$$C_9 = O(N_{\text{FS}} \cdot maxIter \cdot \eta \cdot N_a \cdot N_c \cdot N_o \cdot O(\text{MSDF}^3)) = O\left(N_a^{F+4} \cdot N_o\right) \tag{6.9}$$

where $N_c \leq N_a$, N_{FS} can be upper bounded by N_c^F (in big O notation) and $O(\text{MSDF}^3) = O\left(N_a \log N_a + N_a \cdot \pounds\right)$, where $\pounds(\leq N_a)$ is the average number of successors of an actor.

The complexity of the schedule generation algorithm (Algorithm 13) is calculated as follows. The rank computation for all the uni-processor schedules can be performed in $O(N_s \cdot N_{\text{FS}})$ time, where N_s is the number of uni-processor schedules constructed and N_{FS} is the number of fault-scenarios. The highest throughput rank can be selected in $O(N_s)$ and lines 15–20 can be performed in $O(N_{\text{FS}} \cdot O(\text{SMSDF}^3))$. Finally, the outer while loop (lines 3–25) is repeated N_{FS} times in the worst case. Combining,

$$C_{13} = O\left(N_{\text{FS}} \cdot (N_s \cdot N_{\text{FS}} + N_s + N_{\text{FS}} \cdot O(\text{MSDF}^3))\right)$$
$$= O\left((N_s + O(\text{MSDF}^3)) \cdot N_a^{2F}\right) \tag{6.10}$$

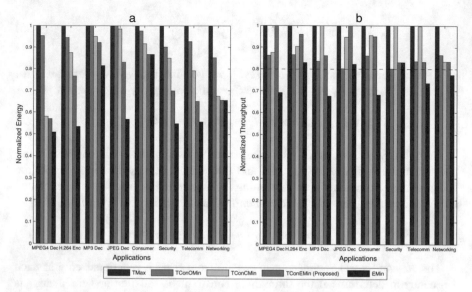

Fig. 6.3 Comparing energy and performance for different applications [5]

6.6.2 Performance of Initial Mapping Selection Algorithm

Figure 6.3 plots the throughput and the energy of our proposed technique compared to three prior studies on reactive fault-tolerance. The starting mapping selection criteria for these works are highest throughput for *TMax* [9], migration overhead minimization with throughput constraint for *TConOMin* [4]), and communication energy minimization with throughput constraint for *TConCMin* [3]), respectively. Additionally, to determine energy overhead incurred in considering throughput in the optimization process, the proposed technique is also compared with the minimum energy starting mapping (*EMin* [7]).

Figure 6.3a plots the normalized total energy consumption per iteration for 8 real-life applications for the existing and the proposed techniques. Energy values are normalized with respect to the *TMax* technique. As can be seen clearly from the figure, energy consumption of the proposed technique (*TConEMin*) is the least among all existing reactive fault-tolerant techniques. On average, for all applications, *TConEMin* achieves 30%, 25%, and 16% less energy as compared to the *TMax, TConOMin*, and *TConCMin*, respectively. The energy savings with respect to *TConCMin* is lower as compared to the other two techniques, because *TConCMin* minimizes communication energy component of the total energy, while the other two techniques do not consider energy optimization. Finally, the *TConE-Min* consumes 15% more energy than *EMin*, which does not consider throughput degradation.

Figure 6.3b plots the normalized throughput of all the techniques. The throughput constraint is shown by the dashed line in the figure. As previously indicated,

Table 6.1 Number of mappings generated in exhaustive search vs our heuristic

	Homogeneous cores	Heterogeneous cores	
Actors	1 type	2 types	3 types
2	2	6	12
4	15	94	309
6	203	2430	12,351
8	4140	89,918	681,870
10	115,975	4,412,798	48,718,569
14	190,899,322	20,732,504,062	461,101,962,108

the *EMin* does not consider throughput degradation, and therefore, throughput constraint is violated for most applications (35 out of 50 applications). Another aspect of the starting mapping generation algorithm is the execution time. The reactive fault-tolerant techniques of [3, 4, 9] search the design space exhaustively to select a starting mapping. Although this is solvable for homogeneous cores with a limited number of actors and/or cores, the problem becomes computationally infeasible, even for small problem size, as the cores become heterogeneous.

Table 6.1 reports the size of the design-space (number of mappings evaluated) as the number of actors increases. The number of cores in the table is same as the number of actors. If the SDF[3] takes an average $10 \, \mu s$ to compute the schedule of a mapping, the design-space exploration time for 14 actors on 14 cores with three types of heterogeneous cores is 54 days. The proposed heuristic solves the optimization problem in less than 2 h, clearly demonstrating the reduction of design space exploration time.

6.6.3 Improvement of Energy Consumption

This section introduces the energy savings obtained during the overall lifetime of an MPSoC as one or more permanent faults occur. Experiments are conducted with the same set of applications as before and executed on an architecture with 6 cores arranged in a mesh architecture of 2×3 organization. For demonstration purpose, the number of permanent faults is limited to 2. These are forced to occur after $n_1 \cdot T$ and $n_2 \cdot T$ years, respectively from the start of device operation, where T is the total lifetime of the device and $0 \leq n_1, n_2 \leq 1$. Figure 6.4 represents the simulation environment. During 0 to $n_1 \cdot T$ years, energy is consumed by the initial mapping, that is, the no-fault mapping discussed in Sect. 6.6.2; during $n_1 \cdot T$ years to $n_2 \cdot T$ years and $n_2 \cdot T$ years to T years, energy is consumed by single fault-scenario and the two fault-scenario mappings, respectively. Cores affected by faults are selected randomly, and results presented here are an average of all single and double faults.

Figure 6.5a plots energy consumption during device lifetime with single fault occurring anytime during its useful life. The average energy consumption per

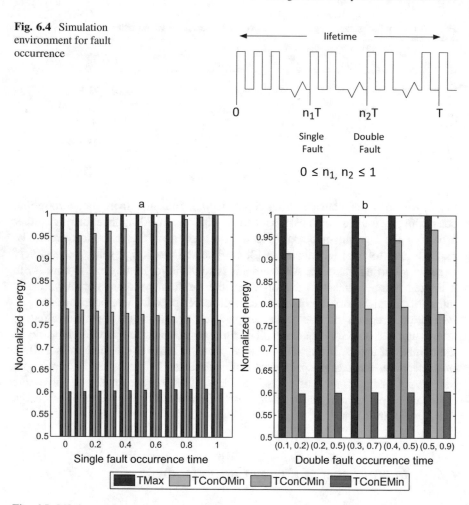

Fig. 6.4 Simulation environment for fault occurrence

Fig. 6.5 Lifetime energy consumption with single and double faults [5]

iteration of the application is plotted with n_1 varied from 0 to 1. A lower value of n_1 implies that the fault occurs in the early life of the device, while a higher value indicates faults occurring after significant usage of the device. As can be seen from this figure, the energy consumption of *TMax* and *TConOMin* techniques is comparable and is higher than that consumed by the other two techniques. This is because these techniques do not optimize computation and communication energy. Although *TConOMin* minimizes migration overhead (energy), this is a one-time overhead (i.e., incurred during fault) and is negligible compared to the total energy consumed in the lifetime of the device. *TConCMin* considers communication energy and throughput jointly, and therefore the energy consumed is lower than *TMax* and *TConOMin* by average 23% and 20%, respectively. Our proposed *TConEMin* achieves an average 22% energy savings compared to *TConCMin*.

Figure 6.5b plots energy results with two faults occurring during the lifetime. A fault-coordinate (n_1, n_2) indicates that the first fault occurs at n_1 and the second at n_2. It is to be noted that n_1, n_2 are in terms of percentage (or fraction) of the overall lifetime. Furthermore, $n_1 \neq n_2$, implying that only one fault occurs at a given time in the device. Although experiments are conducted for all combinations of fault-coordinates, results for a few of the coordinates are plotted. Similar to the single fault results, the proposed *TConEMin* also achieves 30% lower energy as compared to the existing techniques for multiprocessor system with two faults. These results prove that energy-aware mapping selection for fault-tolerance is crucial for minimizing the total energy consumption of a system.

6.6.4   Improvement of Migration Overhead

Table 6.2 reports the migration overhead (measured as energy) and total energy of our proposed technique along with two existing techniques (*OMin* and *TConOMin*) for (*H.264 Encoder* (5 actors) and *MP3 Decoder*) (14 actors). These applications are executed on a multiprocessor system with 6 cores arranged in 2×3. The core heterogeneity is fixed to 2. Columns 3 and 4 report the migration overhead incurred when faults occur and the average energy consumption per iteration of the application graph, respectively. These numbers are average results for 1 and two faults occurring during the lifetime of the platform. Column 5 reports savings in migration overhead achieved by *OMin* and *TConOMin* with respect to our proposed *TConEMin*. Column 5 reports the extra energy (computation + communication) incurred in selecting the same two techniques with respect to *TConEMin*. As can be seen from the table, significant savings in migration overhead are possible with *OMin* technique. However, this technique is associated with an energy penalty (Column 6). For application *H.264 Encoder*, for example, the migration overhead savings in *OMin* is 7×10^8 nJ, while the energy penalty is 3.2×10^5 nJ per iteration. As discussed previously, migration overhead is a one-time overhead while energy is consumed every iteration of the application graph (both pre- and post-fault

Table 6.2 Migration overhead performance

		Migration energy (nJ)	Total energy (nJ)	Migration overhead savings (nJ)	Extra energy per iteration (nJ)	Iterations to recover
H.264 Enc	OMin	1.1×10^9	7.2×10^5	7×10^8	3.2×10^5	2188
	TConOMin	1.7×10^9	4.6×10^5	1×10^8	6×10^4	1667
	TConEMin	1.8×10^9	4.0×10^5	–	–	–
MP3 Dec	OMin	7.0×10^8	2.9×10^6	1.7×10^9	1.4×10^6	1215
	TConOMin	1.3×10^9	2.0×10^6	1.1×10^9	5×10^5	2200
	TConEMin	2.4×10^9	1.5×10^6	–	–	–

occurrence). The savings in migration overhead is compensated in $\frac{7\times10^8}{3.2\times10^5} = 2188$ iterations ($\approx$146 s with a 500 MHz clock at encoding rate of 15 frames per sec). This is reported in column 7.

Results in the last column of the table can also be interpreted as follows. Selecting *TConEMin* as the fault-tolerant technique results in an extra migration overhead of 7×10^8 nJ, which is recovered in the next (post-fault) 2188 iterations of the application graph. For most of the multimedia applications, actors are executed periodically. Examples of these applications on a mobile phone include decoding of frames while playing video and fetching emails from server. Typically, these applications are executed countably infinite times in the entire lifetime of the device. If N denotes the total iterations of a device post-fault occurrence, then the first 2188 iterations will be used to recover the migration overhead loss, while the remaining $(N - 2188)$ iterations will fetch energy savings (3.2×10^5 nJ per iteration). As $N \to \infty$, the energy savings obtained $= (N - 2188) \times 3.2 \times 10^5 \approx N \times 3.2 \times 10^5$ nJ. This substantial energy gain clearly justifies the non-consideration of migration overhead in the fault-tolerant mapping selection.

6.6.5 *Improvement of Throughput*

This section reports results pertaining to the throughput performance of our proposed approach. For this purpose, we broadly classify streaming multimedia applications into two categories—those benefiting from scalable QoS and those requiring a fixed throughput. Majority of the streaming applications, such as video decoding, fall in the latter category. Results presented previously are based on performance (i.e., throughput) as constraint. Those results can be directly attributed to fixed-throughput systems. However, to demonstrate the applicability of our technique for scalable throughput applications, a metric is defined (*throughput per unit energy*). The proposed and the existing techniques are compared based on this metric. Results are presented in Fig. 6.6. Experiments are conducted with a set of six real applications on an architecture with the number of cores varying from two to eight. Core heterogeneity of the architectures is limited to two as the existing techniques fail to provide a solution for the applications with higher core heterogeneity. The results reported in the figure are the average of all single- and double-fault-scenarios. A common trend from these plots is that for most applications (except *H.263 Encoder*), the throughput per unit energy initially increases with the number of cores. On increasing the number of cores beyond the maximum point, the throughput per unit energy decreases. This behavior is the same for all four techniques. This is because, as the number of cores increases, the throughput of an application increases. At the same time, the two energy components (computation and communication) also increase. For lower core count, the growth in throughput dominates, causing an increase in the overall throughput per unit energy. As the core count increases beyond six cores (four cores for *Romberg Integration* and *FFT*), the

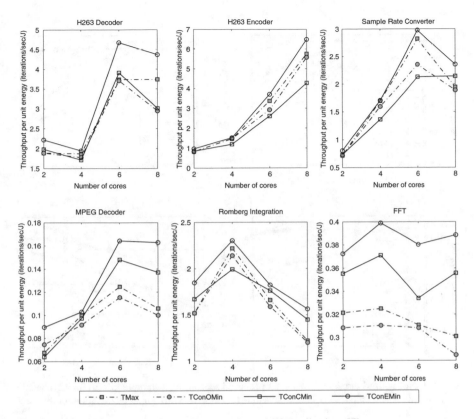

Fig. 6.6 Throughput-energy joint performance for real-life applications [5]

energy growth dominates over throughput growth, and therefore, the throughput per unit energy drops. Although H.263 Encoder shows a growth in throughput per unit energy up to eight cores, the drop-off point is observed for *TConEMin* with 16 cores. However, the results are omitted, as the exhaustive search based existing techniques—*TMax, TConOMin* and *TConCMin* fail to give a solution for this value of core count.

As can be seen, the throughput per unit energy of *TConEMin* is the highest among all existing techniques, delivering on average 30% better throughput per unit energy.

6.6.6 Performance of Our Proposed Scheduling Technique

6.6.6.1 Throughput Results

Figure 6.7a–c plots throughputs obtained in the proposed scheduling technique for six fault-scenarios (three single and three double) while executing *MP3 Decoder*

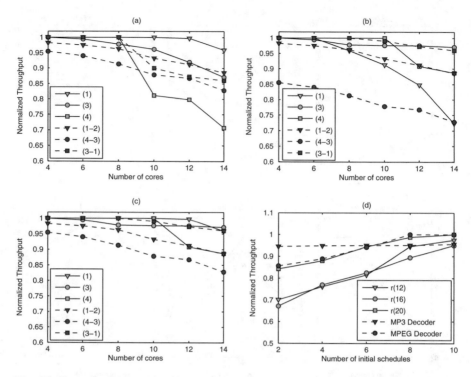

Fig. 6.7 Normalized throughput using our proposed scheduling technique [5]. (**a**) Initial schedule (S_1). (**b**) Initial schedule (S_2). (**c**) Initial schedule (S_1, S_2). (**d**) Throughput degradation for different applications

on a multiprocessor platform where the number of cores is increased from 4 to 14. There are two initial uniprocessor schedules considered ($N_s = 2$). The multiprocessor throughput obtained using these uni-processor schedules is normalized with respect to the throughput obtained using the SDF³ tool and is plotted in Fig. 6.7a, b.

For the initial schedule S_1 in Fig. 6.7a, the normalized throughput for all fault-scenarios decreases with an increase in the number of cores. This is expected, as uniprocessor schedules fail to capture the parallelism available with multiple cores. Among the six fault-scenarios considered, the throughput degradation for fault-scenario *(4)* is the highest ($\approx$30%), while for others, this is less than 20%. Similarly, for Fig. 6.7b (corresponding to initial schedule S_2), fault-scenarios *(1)* and *(4–3)* suffer the highest throughput degradation of 25%. If the two schedules (S_1 and S_2) are considered to be available simultaneously and the one which gives the highest throughput for a fault-scenario is selected as the initial schedule, the throughput degradation can be bounded at design-time. This is shown in Fig. 6.7c, where S_1 is selected as the initial schedule for fault-scenarios *(1)* and *(4–3)* and S_2 as the initial schedule for the remaining fault-scenarios. The maximum throughput degradation obtained using this technique is 18%.

Figure 6.7d plots the throughput degradation obtained as the number of initial schedules is increased from two to ten for five different applications. The results reported in this plot are average results of all single- and double-fault-scenarios. As can be seen from this figure, the throughput degradation decreases with an increase in the number of initial schedules. On average, for all five applications considered, the throughput degradation is within 5% from the throughput constructed using SDF^3 with ten initial schedules. A point to note here is that choosing more initial schedules results in an increase in the storage complexity. Results indicate that $N_s = 10$ (i.e., ten initial schedules) offers the best trade-off with respect to storage and throughput degradation.

6.6.6.2   Overhead Results

Table 6.3 reports the schedule storage overhead (in Kb) and the schedule computation time overhead (in s) using the proposed scheduling technique compared to the storage-based and the construction-based techniques for two applications (*H.263 Encoder* and *MP3 Decoder*) with 5 and 14 actors on an architecture with 12 cores arranged in 3×4. Results are reported for three-fault tolerant systems. The *construction-based* and the proposed technique require storing the fault-tolerant mappings only, while the *storage-based* technique stores the schedule of actors on all cores and for all fault-scenarios alongside the fault-tolerant mappings.

The run-time storage construction overhead is the overhead of fetching a schedule from the database and is negligible compared to constructing a schedule. The *construction-based* technique results in an execution time of 0.4 s. The construction time increases exponentially with the number of actors and/or cores (Column 3 and 6). This large schedule-construction time could potentially lead to deadline violations. The proposed technique results in a linear growth of execution time and is scalable with the number of actors and cores. Finally, the reported execution-time of the design-time analysis phase for the storage-based and the construction-based techniques involves generating schedule for all fault-scenarios. Although schedules are not stored in the construction-based technique, they are still computed at design-time for verification. The corresponding number for the proposed technique denotes the time for constructing initial schedules only. On average, for all 50 applications considered, the proposed technique reduces storage overhead by 10× (92%) with respect to the *storage-based* technique and execution time by 20× (95%) compared to the *construction-based* technique.

Table 6.3 Schedule storage and computation time overheads

Parameters	H.263 Encoder			MP3 Decoder		
	Storage based	Construction based	Proposed	Storage based	Construction based	Proposed
Memory overhead (Kb)	892.1	68.6	68.6	1464	91.5	91.5
Run-time overhead (s)	0	0.42	0.027	0	3.06	0.035
Design-time overhead (s)	34.44	34.44	2.75	80	80	3.66

6.7 Remarks

This work presented a design-time technique to generate mappings of an application on an architecture for all possible core fault-scenarios. The technique minimizes energy consumption while satisfying an application's throughput requirements. Experiments conducted with real and synthetic application SDFGs on a heterogeneous multiprocessor system with different core counts demonstrate that the proposed technique is able to minimize energy consumption by 22%. Additionally, the technique also achieves 30% better throughput per unit energy performance as compared to existing reactive fault-tolerant techniques. A scheduling technique is also proposed based on self-timed execution to minimize schedule construction and storage overhead. Experimental results indicate that the proposed approach achieves 95% less time at run-time for schedule construction. This is crucial in meeting real-time deadlines. Finally, the scheduling technique also minimizes the storage overhead by 92%, which is an important consideration for multimedia applications.

References

1. D. Bertozzi, A. Jalabert, S. Murali, R. Tamhankar, S. Stergiou, L. Benini, G. De Micheli, NoC synthesis flow for customized domain specific multiprocessor systems-on-chip. IEEE Trans. Parallel Distrib. Syst. (TPDS) **16**(2), 113–129 (2005)
2. J. Blazewicz, Scheduling dependent tasks with different arrival times to meet deadlines, in *Proceedings of the International Workshop Organized by the Commision of the European Communities on Modelling and Performance Evaluation of Computer Systems* (North-Holland Publishing Co., 1976), pp. 57–65
3. A. Das, A. Kumar, B. Veeravalli, Energy-aware communication and remapping of tasks for reliable multimedia multiprocessor systems, in *Proceedings of the International Conference on Parallel and Distributed Systems (ICPADS)* (IEEE Computer Society, 2012), pp. 564–571
4. A. Das, A. Kumar, Fault-aware task re-mapping for throughput constrained multimedia applications on NoC-based MPSoCs, in *Proceedings of the International Symposium on Rapid System Prototyping (RSP)* (IEEE, 2012), pp. 149–155
5. A. Das, A. Kumar, B. Veeravalli, Energy-aware task mapping and scheduling for reliable embedded computing systems. ACM Trans. Embed. Comput. Syst. (TECS) **13**(2s), 72:1–72:27 (2014)
6. R. Dick, Embedded System Synthesis Benchmarks Suite (E3S) (2013)
7. J. Hu, R. Marculescu, Energy-aware communication and task scheduling for network-on-chip architectures under real-time constraints, in *Proceedings of the Conference on Design, Automation and Test in Europe(DATE)* (IEEE Computer Society, 2004), p. 10234
8. L. Jiashu, A. Das, A. Kumar, A design flow for partially reconfigurable heterogeneous multi-processor platforms, in *Proceedings of the International Symposium on Rapid System Prototyping (RSP)* (IEEE, 2012) pp. 170–176
9. C. Lee, H. Kim, H.-W. Park, S. Kim, H. Oh, S. Ha, A task remapping technique for reliable multi-core embedded systems, in *Proceedings of the Conference on Hardware/Software Codesign and System Synthesis (CODES+ISSS)* (ACM, 2010), pp. 307–316

10. S. Stuijk, M. Geilen, T. Basten, SDF3: SDF for free, in *Proceedings of the International Conference on Application of Concurrency to System Design (ACSD)* (IEEE Computer Society, 2006), pp. 276–278
11. C. Yang, A. Orailoglu, Predictable execution adaptivity through embedding dynamic reconfigurability into static MPSoC schedules, in *Proceedings of the Conference on Hardware/Software Codesign and System Synthesis (CODES+ISSS)* (ACM, 2007), pp. 15–20

Chapter 7
Run-Time Adaptations for Lifetime Improvement

7.1 Introduction

As discussed in previous chapters, energy consumption and reliability are two important optimization objectives for multiprocessor systems. There is a strong interplay between these two objective functions. Reducing the temperature of a system by efficient thermal management leads to a reduction of leakage power. On the other hand, reduction of power dissipation (by controlling the voltage and frequency of operation) leads to an improvement in the thermal profile of a system. However, too frequent voltage and frequency scaling increases thermal cycles, which increases stress in the metal layers causing reliability concerns. This has attracted significant attention in recent years to investigate on intelligent techniques, such as the use of machine learning, to determine the relationship between temperature, energy and performance, and their control using voltage and frequency switching. State-of-the-art learning-based approaches suffer from the following limitations.

First, multiprocessor systems switch between applications exhibiting performance and workload variations. This changes the thermal behavior of these systems both within (intra) and across (inter) application execution. Although intra-application thermal variations are considered in many recent works, inter-application variations are not addressed. Second, most thermal management techniques focus on average and peak temperature reduction; thermal cycling is not accounted. Last, existing adaptive techniques are either implemented on a simulator or rely on time-consuming thermal prediction using the `HotSpot` tool [7], limiting their accuracy and scalability.

In this work, the above limitations are addressed. We present dynamic thermal management approach for multiprocessor systems that adapts to thermal variations within (intra) and across (inter) application execution. Fundamental to this approach is a run-time system, which interfaces with the on-board thermal sensors and uses reinforcement learning algorithm to learn the relationship between mapping of

A.K. Das et al., *Reliable and Energy Efficient Streaming Multiprocessor Systems*, Embedded Systems, https://doi.org/10.1007/978-3-319-69374-3_7

threads to cores, the voltage-frequency of a core, and its temperature. The aim is to control three thermal parameters (1) the peak temperature, (2) the average temperature, and (3) the thermal cycling. The objective is to achieve an extended lifetime of these systems, measured in terms of mean time to failure (MTTF). This work makes the following contributions:

1. inter-and intra-application thermal management using thread-to-core allocation (using CPU affinity[1]) and dynamic frequency control (using CPU governors[2]);
2. separation of temperature sampling interval from the decision interval of conventional reinforcement learning algorithm to accurately model (and hence control) average temperature and thermal cycling; and
3. implementation of the run-time system incorporating the machine learning algorithm on a real platform.

The proposed approach is implemented on an Intel quad-core platform running Linux kernel 3.8.0. A set of multimedia applications from the *ALPBench* suite [4] are executed on the platform. Results demonstrate that the proposed approach minimizes average temperature and thermal cycling, leading to a significant improvement in MTTF as demonstrated in Sect. 7.4. Additionally, the static and the dynamic energy consumption are reduced by 11% and 10%, respectively.

The remainder of this chapter is organized as follows. Overview of the proposed reinforcement learning-based thermal management approach is presented in Sect. 7.2 and the implementation details in Sect. 7.3. Results of the proposed technique are presented next in Sect. 7.4, and the chapter is concluded in Sect. 7.5.

7.2 Proposed Thermal Management Approach

Figure 7.1a visualizes the three design layers typical of a modern multiprocessor system. A general overview is provided on the interaction of these three layers.

Application Layer The application layer is composed of a set of applications, which are executed on the system. Every application is characterized by a performance requirement, which determines the quality-of-service for the corresponding application. Table 7.1 reports the performance metric for some of the common applications for modern multiprocessor systems. An application is annotated to include its timing requirement, which is communicated to the operating system through the application programming interface. Additionally, the source code of the application is modified to insert breakpoints in order to signal the operating system to recalculate the thread affinities and the voltage-frequency values for the next execution interval. For video/image applications, breakpoints are inserted

---

[1]CPU affinity enables the binding of an application thread to a physical core or a range of cores.
[2]CPU governors are power schemes for the CPU, deciding the frequency of operation of the cores.

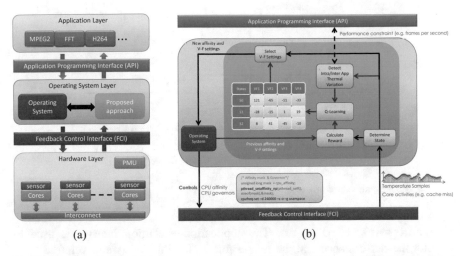

Fig. 7.1 Proposed thermal management approach. (**a**) Three design layers. (**b**) Q-learning based thermal optimization

Table 7.1 Performance metric for commonly executed multiprocessor applications

Applications	Performance metric
MPEG2, MPEG4, H264	Time to encode/decode a video frame
JPEG Enc/Dec	Time to encode/decode an image
FFT/iFFT	Time for 256 Fourier transforms
AES, SHA	Time to hash 2048-byte message
Basicmath, gzip, bitcount	Time for 100 operations

after encoding or decoding of every frame. Similarly for FFT/iFFT, the breakpoint interval is every 256 Fourier transforms. Figure 7.2 shows an example breakpoint-based program execution. The timing overhead τ_o incorporates the following: the time for the Q-learning algorithm to generate the new thread affinities and voltage-frequency values; time to migrate the application threads; and the time to set the voltage and frequency on the CPU cores through the operating system.

Operating System Layer The operating system layer is responsible for coordinating the application execution on the hardware. Of the different responsibilities of the operating system, such as scheduling, memory management, and device management, the focus here is on the cross-layer interaction, which forms the background of the proposed machine learning-based thermal management. At every application breakpoint, the operating system stalls the application execution and triggers the proposed machine learning algorithm, which generates the new thread affinities and voltage-frequency values for the CPU cores. The operating system applies the new voltage-frequency values on the CPU cores using the `cpufreq-set` utility. The application threads are migrated to the corresponding cores, as specified using the affinity masks, using the operating system command

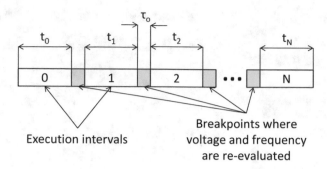

Fig. 7.2 Breakpoints-based task execution

`pthread_setaffinity_np`. The application execution interval is used to monitor the performance and the temperature. This interval is also referred to as *decision epoch* in the machine learning terminology. Throughout the rest of this work, the execution interval and the *decision epoch* are used interchangeably.

In most existing works on Q-learning based thermal management, the decision epoch is used to sample the temperature; actions are selected based on the instantaneous temperature from the sensor, which is not a true indication of the average temperature or thermal cycling in the interval. Since temperature-related reliability is governed by average temperature and thermal cycling, which need to be measured over a period of time, the temperature sampling interval is lower than of the decision epoch for this work.

Hardware Layer The hardware layer consists of the processing cores with thermal sensors and performance monitoring unit (PMU) to record the performance statistics. Of the different performance statistics available, this work considers *CPU cycles*, which gives a fair indication of the workload of a given application. The temperature samples are collected continuously by the operating system at the temperature sampling interval. The PMU readings are collected at every breakpoint and subsequently, the readings are reset to allow performance recording for the next execution interval. Finally, before the start of the next execution interval, the following sequence of events takes place:

- the application threads are migrated to the corresponding cores as specified using the `pthread_setaffinity_np` command;
- the frequency value, set by the operating system using the `cpufreq-set` command, is converted to a corresponding CPU clock divider setting;
- the divider settings are written into appropriate CPU registers; and
- the divided CPU clock is used to execute the migrated thread for the next execution interval.

7.3   Q-Learning Implementation

The Q-learning [10] based thermal management approach is integrated in the operating system layer as discussed before. To give more insight into the approach, Fig. 7.1b shows its internal details. Different components of the proposed approach are discussed next starting with a general overview of the Q-learning algorithm.

7.3.1   General Overview

The Q-learning is a reinforcement learning technique used to find the optimum policy of a given Markov Decision Process (MDP). A simplistic view of the Q-learning is shown in Fig. 7.3. The algorithm consists of a learning agent that works by observing the state (s) of the environment and selecting a suitable action (a) to control the state. The learning is quantified and stored in a table (referred to as Q-Table in machine learning terminology) corresponding to every state-action pair. Every entry of this table, corresponding to the state-action pair (s, a), represents the reward (or penalty, if the entry is negative) obtained by selecting action a when the environment is in state s. The algorithm works by always selecting the action with the highest reward for any observed state of the environment.

In the context of multiprocessor thermal management, the learning steps are:

S1:   determine the previous state-action pair;
S2:   update the Q-Table entry for this state-action pair based on the current state;
S3:   select thread affinities and voltage-frequencies based on the current state;
S4:   apply the selected action for the next execution interval.

Referring back to Fig. 7.1b, step **S1** is implemented in the *Calculate Reward* block, step **S2** in the *Q-learning* block, step **S3** is implemented in the *Select V-F Settings* block, and finally step **S4** in the *Operating System* block. The block *Detect Intra/Inter App Thermal Variation* is used to update the Q-Table.

Fig. 7.3 Basic Q-learning

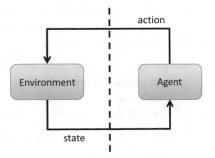

7.3.2 Determine State

As established in Sect. 7.2, the Q-learning based thermal management is triggered at every application breakpoint. The *Determine State* block of the approach determines the current state of the system from the temperature readings of the thermal sensors. These temperature readings are used to calculate the thermal *aging* $\mathscr{A}$ (refer to Eq. 2.21) and thermal *stress* $\mathscr{S}$ (refer to Eq. 2.11). To limit state space explosion, the working range of these parameters is divided into N_a and N_s disjoint intervals, respectively. Specifically, *stress* is the set $\mathscr{S} = \{(0, s_0], (s_0, s_1], \cdots, (s_{N_s-1}, s_{N_s}]\}$ and the symbol $\hat{s}_i$ is used to represent the interval $(s_i, s_{i+1}]$. Similarly, *aging* is the set $\mathscr{A} = \{(0, a_0], (a_0, a_1], \cdots, (a_{N_a-1}, a_{N_a}]\}$ and the symbol $\hat{a}_i$ is used to represent the interval $(a_i, a_{i+1}]$. The environment is represented as $\mathscr{E} : (\mathscr{A} \times \mathscr{S})$.

7.3.3 Action Space of the Q-Learning Algorithm

The thermal behavior of a multiprocessor system is influenced by CPU affinity and the voltage-frequency settings. These form the action space of the Q-learning algorithm, which is represented by $A : (\mathscr{M} \times \mathscr{V})$ where $\mathscr{M}$ is the set of thread affinity mappings and $\mathscr{V}$ is the set of governors.

7.3.4 Q-Learning Algorithm

The Q-learning block in Fig. 7.1b is responsible for updating the Q-Table entries. The reward or penalty for selecting an action at a state is calculated at the next breakpoint based on the goodness of the selected action. In other words, at each breakpoint, the state-action pair of the last execution interval is evaluated. For this, the last selected action is read in a local variable `laction`, through the operating system commands `cpufreq-info` and `pthread_getaffinity_np`. The last state is stored in another local variable `lstate`. This variable is updated with the current state, once its old value is used to update the Q-Table entry. The goodness of the last action in the last state is measured as reward (or penalty if it is negative) and is calculated based on

- *the performance slack*, i.e., the difference between the time required to execute the last execution interval and the timing requirement specified in the annotated application source code; and
- *the thermal safety*, i.e., the current thermal state of the system.

Mathematically, this is represented as

$$r = \begin{cases} -\hat{s}_i \times \hat{a}_i & \text{if } (\hat{s}_i = \hat{s}_{N_s}) \text{ or } (\hat{a}_i = \hat{a}_{N_a}) \\ f(\hat{a}_i, \hat{s}_i) + (P_c - P) & \text{otherwise} \end{cases} \qquad (7.1)$$

where P is the performance, P_c is the performance constraint, and the function f is determined empirically as $f = (a.K_1.stress + b.K_2.aging)$, where a and b are relative importance of *stress* and *aging*: For mpeg (large thermal cycles), $a > b$ and for tachyon (high average temperature), $b > a$. Two sets of a and b values are used based on the mean of *stress* and *aging*. $K_1(K_2)$ is the learning weight and is a Gaussian function of the *stress* (*aging*) values. This distribution assigns lower rewards to thermally unstable as well as the thermal stable states and thus, allows the algorithm to explore other states and prevent Q-Table clustering.

For the design of the reward function, two cases are considered. If the *stress* or *aging* falls in the unsafe zone (the last interval), the decision is penalized. This is indicated with a negative value of the reward function. For all other cases, the reward function is composed of performance penalty and the thermal safety of the state. Specifically, if the performance requirement is not satisfied, $(P_c - P)$ is negative and the reward (or penalty) is governed by the function f. Finally, rewards are guaranteed if an action leads to a thermal safe state while satisfying the performance requirements. The reward value of Eq. 7.1 is used to update the Q-table entry using the following equation:

$$Q(\texttt{lstate}, \texttt{laction}) = Q(\texttt{lstate}, \texttt{laction}) + \alpha \cdot r_i \qquad (7.2)$$

where α is the learning rate. One of the interesting features of the Q-learning algorithm is its three phases of operation. This is shown in Fig. 7.4. These three phases are *exploration*, *exploration-exploitation*, and *exploitation*. In the *exploration phase*, the algorithm learns the goodness of all the different actions. Therefore, the table entries are updated with all of the calculated reward values (Eq. 7.1). This phase is characterized by $\alpha = 1$. On the other hand, in the *exploitation* phase, the algorithm selects the action based on the previous learning. Therefore, the table entries are not updated with the reward value. This phase is characterized by $\alpha = 0$. In between these two phases, the algorithm remains in the *exploration-exploitation* phase. In this phase, the algorithm learns the goodness of only the good actions (outcome of the *exploration* phase) to evaluate it further. This phase is therefore characterized by $0 < \alpha < 1$.

Fig. 7.4 Three phases of the Q-learning algorithm

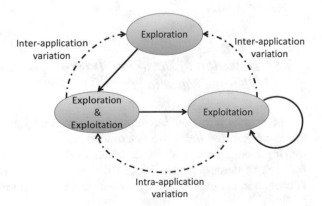

Fig. 7.5 Change of α with
number of breakpoints

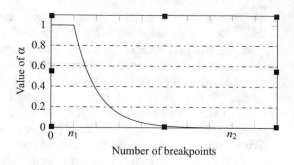

Figure 7.5 plots the change in the alpha values for these three phases against the number of invocations of the Q-learning algorithm, i.e. the number of breakpoints. The *exploration* phase is active for 0 to n_1 breakpoints, the *exploration-exploitation* phase is active for n_1 to n_2 breakpoints and the *exploitation* phase is active beyond n_2 breakpoints. Let N denote the total number of breakpoints to be inserted in the application execution (refer to Fig. 7.2). The following lemma can be stated.

Lemma 1 *The convergence of any Q-learning algorithm is guaranteed if $N \geq n_2$.* In the context of this work, the above lemma states that for thermal management using Q-learning algorithm, the number of breakpoints to be inserted in the application execution should be such that the algorithm reaches the *exploitation* phase. However, too many breakpoints during a small application execution will incur performance penalty. The performance parameters in Table 7.1 are defined based on these considerations.

The state transition diagram of the unmodified Q-learning is shown in Fig. 7.4 with the dark solid lines. To introduce autonomous reaction of the algorithm to intra- and inter-application thermal variations, the state transition diagram is modified using the dotted lines as shown in the figure. When an intra-application thermal variation is detected, the modified algorithm transits to the *exploration-exploitation* phase and the α value is changed accordingly. On the other hand, for inter-application variation, the modified algorithm transits to the *exploration* phase resetting $\alpha = 1$. The detection of the intra- and inter-application thermal variations is discussed next.

7.3.5   Detection of Intra- and Inter-Application Thermal Variations

To incorporate intra- and inter-application thermal variations, moving averages of the *stress* and the *aging* are determined at every breakpoint. The change in the moving averages are identified as ΔMA_s and ΔMA_a. Two thresholds are maintained for each of these quantities identified with the superscript L and U, respectively. A change in the moving average is considered as intra-application variation, if the change is greater than the lower threshold and lower than the upper threshold (for example, when $\Delta MA_s^L \leq \Delta MA_s < \Delta MA_s^U$). On the other hand, the change in the moving average is considered as inter-application variation, if the change is greater than the upper threshold.

7.4 Results

Our proposed run-time approach is validated experimentally on an Intel quad-core CPU running Linux kernel 3.8.0. Performance is monitored using `perf` [6]; temperature is monitored by sampling thermal sensors; power/energy consumption is recorded using `likwid-powermeter` [8]. A set of multi-threaded multimedia applications are considered from the `ALPBench` [4] benchmark suite. These benchmarks are `mpeg enc`, `mpeg dec`, `face recognition`, `sphinx`, and `tachyon`. These are representative of the multimedia workloads for most multiprocessor systems. The number of threads in each of these applications is configurable. For our experiments, the number of threads is set to six. Material parameters for computing `aging` and `stress` of a core are the same as that used in [1, 9].

7.4.1 Intra-Application

Table 7.2 reports the average and peak temperatures obtained using three techniques—our proposed approach, Linux's ondemand [5] governor, and the thermal management technique proposed in [3]. Results are reported for three different applications, each of which is executed for three sets of input data. The

Table 7.2 Average and peak temperature (°C) of the proposed approach with respect to state-of-the-art

Benchmarks		Average temperature (°C)			Peak temperature (°C)		
Application	Data	Linux [5]	Ge et al. [3]	Proposed	Linux [5]	Ge et al. [3]	Proposed
tachyon	Set 1	69.2	52.6	50.6	71.5	63.0	60.0
	Set 2	50.5	44.5	43.8	57.3	56.3	52.0
	Set 3	50.8	44.7	41.6	57.8	54.5	48.8
mpeg dec	Clip 1	36.0	34.0	34.2	42.7	41.3	39.0
	Clip 2	35.6	34.4	34.2	42.3	42.0	39.3
	Clip 3	34.3	34.4	34.0	43.0	39.7	44.3
mpeg enc	Seq 1	33.7	34.1	32.6	41.0	40.7	40.3
	Seq 2	34.4	33.5	32.3	41.3	39.7	41.7
	Seq 3	33.2	33.7	31.8	40.3	40.0	41.0

reliability measured as mean time to failure (MTTF) due to these three approaches and data set are reported in Table 7.3. Specifically, the table reports three MTTF values—(1) MTTF due to thermal cycling, (2) MTTF due to thermal aging, and (3) the minimum of the two MTTF values.

Few observations can be made from the results presented in this table. First, the technique of Ge et al. [3] minimizes instantaneous temperature achieving a lower aging (i.e., higher thermal aging related MTTF) than Linux's Ondemand, which does not consider thermal aspects. This can be seen in Table 7.2 columns 3–4 and Table 7.3 columns 6–7. These results highlight the importance of thermal management feature for operating systems in effectively maximizing the thermal reliability. However, thermal cycling is not accounted for in this technique and therefore does not guarantee reduction of thermal stress. This is evident from the thermal cycling-related MTTF values in Table 7.3 columns 3–4 for scenarios such as tachyon on set 1 and mpeg dec on clip 1, where MTTF values obtained using [3] are lower than that of Linux Ondemand governor.

Second, our proposed reinforcement learning-based approach minimizes the average temperature by up to 18.6°C and the peak temperature by up to 11.5°C. This reduction leads to improvement in aging-related MTTF by up to 5× (average 82%) as reported in Table 7.3 column 8. Third, the proposed approach also minimizes thermal cycling which is not considered in Linux's Ondemand governor and in the approach of Ge et al. [3]. This can be seen in Table 7.3 column 5. It is to be noted that, for the tachyon application with set 1 data, the thermal cycling-related MTTF using Linux's default thread assignment is higher (7.1 years); however, the aging-related MTTF is lower (0.7 years). Our proposed technique balances the two effects and improves aging-related MTTF by 5× with less than 25% sacrifice in thermal cycling-related MTTF (while still maintaining a satisfactory MTTF of 5.5 years). For all other applications and data sets, our proposed approach outperforms Linux's Ondemand governor in terms of thermal cycling by an average 2.3×.

Last, our proposed approach outperforms the thermal management technique of Ge et al. [3] both in terms of aging (an average 13% higher aging related MTTF) and thermal cycling (an average 2× higher thermal cycling related MTTF). While the improvement of thermal cycling is expected (as this is incorporated explicitly in the proposed approach), the improvement in aging is due to the following. First, decoupling of the temperature sampling interval from the decision epoch (enabling a finer control on the average temperature); second, careful choice of the design parameters as demonstrated in Sect. 7.4.3.

7.4.2 Inter-Application

Figure 7.6 plots the normalized MTTF due to thermal cycling obtained using the proposed technique compared to that obtained using the modified technique of [3] (referred in this figure as Modified Ge et al.) for six different inter-application scenarios. The MTTF values are normalized with respect to the MTTF obtained

Table 7.3 MTTF (in years) of reinforcement learning algorithm for three applications

Benchmarks		Thermal cycling MTTF			Thermal aging MTTF			Overall MTTF		
Application	Data	Linux [5]	Ge et al. [3]	Proposed	Linux [5]	Ge et al. [3]	Proposed	Linux [5]	Ge et al. [3]	Proposed
tachyon	Set 1	7.1	2.3	5.5	0.7	3.0	3.6	0.7	2.3	3.6
	Set 2	2.8	4.3	5.3	2.6	4.5	4.8	2.6	4.3	4.8
	Set 3	1.3	3.8	6.5	2.4	4.1	5.5	1.3	3.8	5.5
mpeg dec	Clip 1	2.1	0.8	6.4	3.7	4.5	4.4	2.1	0.8	4.4
	Clip 2	1.1	0.9	4.7	3.8	4.3	4.4	1.1	0.9	4.4
	Clip 3	1.6	3.4	3.7	4.3	4.2	4.5	1.6	3.4	3.7
mpeg enc	Seq 1	4.3	4.4	5.2	4.6	4.5	5.2	4.3	4.4	5.2
	Seq 2	3.9	6.2	4.8	4.3	4.7	5.4	3.9	4.7	4.8
	Seq 3	4.6	5.1	5.1	4.9	4.6	5.7	4.6	4.6	5.1

The scaling parameters for computing MTTF are so selected such that the MTTF of an unstressed core (i.e., an idle core) is 10 years

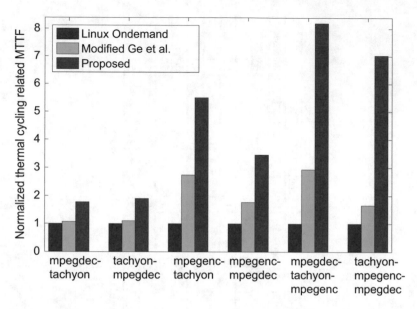

Fig. 7.6 Inter-application results, showing a set of 4 two-application scenarios and 2 three-application scenarios [2]

using Linux's Ondemand governor. Furthermore, the technique of [3] is modified to consider application switching using explicit indication from the application layer. The proposed approach, however, detects application switching autonomously (without communication from the application layer) and performs re-learning.

There are six inter-application scenarios considered in this experiment. A scenario A-B indicate that A is executed first followed by application B. As can be clearly seen from the figure, the technique of Ge et al. [3] results in higher MTTF than Linux's Ondemand governor. For some inter-application scenarios such as mpegdec-tachyon and tachyon-mpegdec, improvements are less ($\approx$8%). For other inter-application scenarios, improvements are higher. On average, the technique of [3] increases MTTF by 80% compared to Linux's Ondemand governor. Our approach outperforms both Linux's Ondemand governor and that of [3] in terms of thermal cycling, achieving 5× and 3× improvements, respectively.

7.4.3 Convergence of Q-Learning Algorithm

Figure 7.7 plots the convergence time of the proposed algorithm for the *mpeg decoding* application with a varying number of states and actions. The convergence time is measured by the number of decision epochs needed to train the proposed learning algorithm. As can be seen from the figure, the number of iterations, i.e., the training time increases with an increase in the number of actions and states. This is

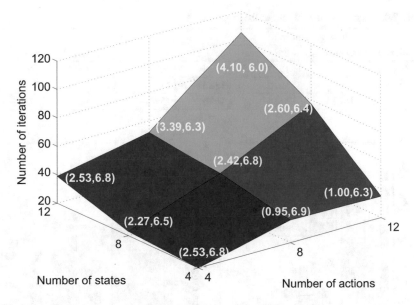

Fig. 7.7   Convergence of the reinforcement learning algorithm [2]

expected because, an increase in the number of states or actions leads to an increase in the size of the Q-Table and therefore more iterations are needed to learn (populate the table entries). The figure also reports the MTTF as coordinates (thermal cycling, aging) for each design point. As the size of the Q-Table increases, the reinforcement learning algorithm has finer control on the temperature, resulting in an improvement in the MTTF. The number of states and actions are chosen based on this learning time and solution quality trade-off.

7.4.4   Execution Time of the Proposed Approach

Table 7.4 reports execution time of the proposed approach compared to the one proposed in [3] and the Linux ondemand, powersave, and user-space power governors. Two user frequencies (2.4 GHz and 3.4 GHz) are shown in the table. Execution time with the highest frequency of 3.4 GHz is the least for all the applications. This is expected because execution time is inversely proportional to operating frequency. For the same reason, execution time with the lowest frequency (powersave) is the highest. For some applications such as tachyon, the proposed approach has higher execution time than the Linux's ondemand governor by up to 30%. This is because the workload in tachyon forces the Linux Kernel to execute always at the highest frequency of 3.4 GHz in the ondemand mode. Thus, the execution time for ondemand and *userspace-3.4* GHz is comparable. The

Table 7.4 Execution time (in sec) of the proposed approach

Application	Linux governors				Ge et al. [3]	Proposed
	Ondemand	Powersave	2.4 GHz	3.4 GHz		
tachyon	629	1337	913	627	1137	810
mpeg dec	1208	1222	1183	1127	1328	1204
mpeg enc	1623	1655	1628	1571	1676	1599

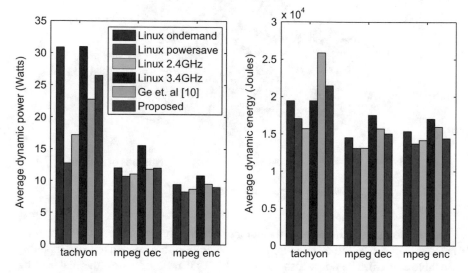

Fig. 7.8 Power comparison of the reinforcement learning algorithm [2]

proposed approach, on the other hand, explores different power modes to reduce thermal thermal cycling and aging and therefore trades off performance. For other applications such as the mpeg enc and the mpeg dec, the execution time of the proposed approach is lower than that of the *ondemand* governor. Finally, with respect to [3], the proposed approach reduces execution time by an average 14%.

7.4.5   Power Consumption of the Proposed Approach

Figure 7.8 plots the average dynamic power and energy consumption (measured using likwid-powermeter) of the proposed approach compared to that of [3] and the Linux governors. Although the power and energy overhead are not reported in [3], these are measured and presented here for a comparative study. As can be seen from the figure, our proposed approach reduces power consumption by an average 6% compared to Linux Ondemand governor with 10% increase in execution time. Although the dynamic power consumption of [3] is lower than the proposed approach (on average 4% lower), the energy consumption (which incorporates both power and execution time) of the proposed approach is 10% lower than that of [3] and within 3% of the energy consumption of Linux's *ondemand* governor. It is to be noted that, by reducing the average temperature, the proposed technique improves leakage power. Using the computation principle reported in [9], our approach improves leakage energy by an average 15% compared to the Linux ondemand governor and 11% as compared to [3].

7.5 Remarks

In this chapter, a reinforcement learning-based run-time approach is proposed for multiprocessor system to adapt to thermal variations both within an application and when the system switches from one application to another. The control is provided by overriding the operating system mapping decisions using affinity masks and dynamically changing the frequency of cores using CPU governors. The approach is validated experimentally using an Intel quad-core platform running Linux. Results demonstrate that the proposed approach improves MTTF by an average 2× for intra-application and 3× for inter-application scenarios as compared to the existing dynamic thermal management technique. Furthermore, the approach also improves dynamic energy consumption by an average 10% and static energy by 11%.

References

1. T. Chantem, Y. Xiang, X.S. Hu, R.P. Dick, Enhancing multicore reliability through wear compensation in online assignment and scheduling. in *Proceedings of the Conference on Design, Automation and Test in Europe (DATE)* (European Design and Automation Association, 2013), pp. 1373–1378
2. A. Das, R.A. Shafik, G.V. Merrett, B.M. Al-Hashimi, A. Kumar, B. Veeravalli, Reinforcement learning-based inter- and intra-application thermal optimization for lifetime improvement of multicore systems, in *Proceeding of the Annual Design Automation Conference (DAC)* (ACM, 2014)
3. Y. Ge, Q. Qiu, Dynamic thermal management for multimedia applications using machine learning, in *Proceeding of the Annual Design Automation Conference (DAC)* (ACM, 2011), pp. 95–100
4. M.-L. Li, R. Sasanka, S.V. Adve, Y.-K. Chen, E. Debes, The ALPBench benchmark suite for complex multimedia applications, in *Workload Characterization Symposium* (IEEE, 2005), pp. 34–45
5. V. Pallipadi, A. Starikovskiy, The ondemand governor, in *Proceedings of the Linux Symposium*, vol. 2 (2006), pp. 215–230
6. Perf: Linux Profiling with Performance Counters (2012), https://perf.wiki.kernel.org
7. K. Skadron, M.R. Stan, K. Sankaranarayanan, W. Huang, S. Velusamy, D. Tarjan, Temperature-aware microarchitecture: modeling and implementation, ACM Trans. Archit. Code Optim. (TACO) **1**(1), 94–125 (2004)

8. J. Treibig, G. Hager, G. Wellein, LIKWID: a lightweight performance-oriented tool suite for x86 multicore environments, in *International Conference on Parallel Processing Workshops (ICPPW)* (2010), pp. 207–216
9. I. Ukhov, M. Bao, P. Eles, Z. Peng, Steady-state dynamic temperature analysis and reliability optimization for embedded multiprocessor systems, in *Proceeding of the Annual Design Automation Conference (DAC)* (ACM, 2012), pp. 197–204
10. C. Watkins, P. Dayan, Q-learning. Mach. Learn. **8**(3–4), 279–292 (1992)

Chapter 8
Conclusions and Future Directions

Reliability and energy are emerging as two of the growing concerns for multiprocessor systems at deep sub-micron technology nodes. This book presented a system-level approach, namely application mapping and scheduling, to jointly address the reliability and energy problems for multiprocessor systems. A detailed background is presented in Chap. 3 on synchronous data flow graphs (SDFGs), used as application models in this book. This chapter also presented the mathematical background on lifetime reliability and the related studies on task mapping and scheduling for lifetime improvement.

In Chap. 4, a platform-based design methodology is proposed that involves task mapping on a given multiprocessor system to jointly minimize energy consumption and temperature-related wear-out, while satisfying the performance requirement. Fundamental to this methodology is a simplified temperature model that incorporates not only the transient and steady-state behavior (temporal effect), but also its dependency on the temperature of the surrounding cores (spatial effect). The proposed temperature model is integrated in a gradient-based fast heuristic that controls the voltage and frequency of the cores to limit the average and peak temperature leading to a longer lifetime, simultaneously minimizing the energy consumption. A linear programming approach is then proposed to distribute the cores of a multiprocessor system among concurrent applications (use-cases) to maximize the lifetime. Experiments conducted with a set of synthetic and real-life applications represented as SDFGs demonstrate that the proposed approach minimizes energy consumption by an average 24% with 47% increase in lifetime.

In Chap. 5 of this book, a fast design space exploration technique is presented for hardware–software partitioning by analyzing the negative impact of increasing the number of checkpoints for transient fault-tolerance on the lifetime reliability of the processing cores. Based on this, a hardware–software co-design approach is proposed to determine the architecture of a reconfigurable multiprocessor system to maximize its lifetime reliability by considering applications enabled individually and concurrently. Experiments conducted with real life and synthetic applications

© Springer International Publishing AG 2018
A.K. Das et al., *Reliable and Energy Efficient Streaming Multiprocessor Systems*,
Embedded Systems, https://doi.org/10.1007/978-3-319-69374-3_8

represented as SDFGs on a reconfigurable multiprocessor system demonstrate that the proposed technique improves lifetime reliability by an average 65% for single applications and an average 70% for use-cases with 25% fewer GPPs and 20% less reconfigurable area as compared to the existing hardware–software co-design approaches.

To ensure graceful performance degradation in the presence of faults, a design-time (off-line) multi-criterion optimization technique is proposed in Chap. 6 for application mapping on embedded multiprocessor systems to minimize energy consumption for all processor fault-scenarios. A scheduling technique is then proposed based on self-timed execution to minimize the schedule storage and construction overhead at run-time. Experiments conducted with SDFGs on hetero-geneous multiprocessor systems demonstrate that the proposed technique minimizes energy consumption by 22% and design space exploration time by 100×, while satisfying the throughput requirement for all processor fault-scenarios. For scalable throughput applications, the proposed technique achieves 30% better throughput per unit energy, compared to the existing techniques. Additionally, the self-timed execution-based scheduling technique minimizes schedule construction time by 95% and storage overhead by 92%.

Finally, an adaptive thermal management approach is proposed in Chap. 7 as a part of the run-time methodology, to improve the lifetime reliability of multiprocessor systems by considering both inter- and intra-application thermal variations. Fundamental to this approach is a reinforcement learning algorithm, which learns the relationship between the mapping of threads to cores, the fre-quency of a core and its temperature (sampled from on-board thermal sensors). Action is provided by overriding the operating system's mapping decisions using affinity masks and dynamically changing CPU frequency using in-kernel governors. Lifetime improvement is achieved by controlling not only the peak and average temperatures, but also thermal cycling, which is an emerging wear-out concern in modern systems. The proposed approach is validated experimentally using an Intel quad-core platform executing a diverse set of multimedia benchmarks. Results demonstrate that the proposed approach minimizes average temperature, peak temperature, and thermal cycling, improving the mean time to failure by an average of 2× for intra-application and 3× for inter-application scenarios when compared to existing thermal management techniques. Additionally, the dynamic and static energy consumption are also reduced by an average 10% and 11%, respectively.

8.1   Near and Far Future Challenges

While this book presented the design methodologies for multiprocessor systems to jointly optimize reliability and energy consumption, a number of issues remain to be solved. Some of these are highlighted below.

8.1.1 Future Challenges for Design-Time Analysis

Heterogeneity Challenges One of the key challenges associated with heterogeneous architectures involves task mapping. The execution times of a task on different processors are different. Further, not every task can be executed on every processor. This makes the task mapping and scheduling problem difficult to solve. The task migration approach also needs to be re-visited. This is because the new object code of a task needs to be compiled and transferred to the migrating core, if this is of different type than the core where the task was executing initially. Storing the pre-compiled object code of all tasks for all core type on every core incurs significant storage overhead and is crucial for multimedia multiprocessor systems where storage space is limited. Further, the transfer of object code on the networks-on-chip could lead to potential congestion of the same and could interfere with the data communication among the non-faulty cores. These challenges need to be solved for the existing design-time based approaches.

Dynamic Partial Reconfiguration This book has shown that there exists a significant scope of reliability and energy improvement for FPGA-based reconfigurable multiprocessor systems. Although reconfiguration of the FPGA logic is allowed before enabling an application on the system, it is anticipated that further improvement is possible by allowing dynamic partial reconfiguration, i.e., by allowing a part of the FPGA logic to be programmed for an application, while the remaining part of the logic is in operation. This will allow to partition more tasks as hardware tasks (i.e., tasks that will be executed on the FPGA logic) during the hardware–software partitioning stage. This will reduce the stress on the processing cores, increasing their lifetime reliability. However, dynamic partial reconfiguration involves timing overhead to program the FPGA bits. Thus, complex analysis framework needs to be developed to predict the performance and improve lifetime reliability. Additionally, dynamic voltage and frequency scaling capabilities are to be explored, which is not addressed currently for reconfigurable multiprocessor systems.

3D Multiprocessor System Most of the existing reliability optimization techniques are limited to 2D multiprocessor systems. Recent developments on 3D systems have attracted a significant research attention towards resource allocation and management for 3D systems. These systems offer significant performance and energy advantages over the native 2D multiprocessor systems, but is known to suffer from thermal issues. An effective approach to solve this challenge involves a thorough understanding of the reliability concerns for these systems, including the detailed transistor-level reliability modeling, which is still at an early stage of development; the thermal characterization of 3D systems; and finally the resource allocation for reliability improvement.

8.1.2   Future Challenges for Run-Time Analysis

The proposed run-time thermal management for multiprocessor systems addressed intra- and inter-application workload variations. The next step towards this is to solve the following three key challenges—accuracy, scalability, and scope.

Accuracy   Modern commercial off-the-self multiprocessor systems, such as Intel Ivy Bridge or Haswell architectures, are equipped with cooling mechanisms in the form of heat sink and fan. The temperature of the different cores of the architecture varies according to the fan speed and the heat sink mechanism. The temperature is also dependent on the placement of these cooling mechanisms in the floorplan. Specifically, cores located closer to the heat sink are often cooler than other cores, even when all the cores execute the same workload. Additionally, the speed of the fan can also be controlled by controlling the current drawn and therefore, the heat dissipation can be adapted based on the application workload. An efficient thermal management needs to incorporate (1) the floorplan information and (2) workload-aware cooling control, alongside the existing techniques, such as dynamic voltage and frequency scaling (DVFS) and task mapping.

Scalability   The future of multiprocessor systems is many-core architecture, integrating hundreds of cores on the same die. The shared memory architecture for multi-core system is expected to become a bottleneck for many-core systems, and therefore message passing interface (MPI) is expected to become the standard communication protocol for many-core architectures. The existing techniques for multi-core systems need to be re-visited from the MPI perspective. Another challenge in the many-core domain is concerning the scalability of the algorithm itself. The problem of finding an optimal task mapping that reduces the peak or average temperature is an NP-hard problem. This problem grows by several orders of magnitude for many-core architecture. Moreover, specific optimization technique also need to be adapted to address the scalability issues. As an example, for the machine learning-based thermal management, the centralized learning model will become a bottleneck for many-core architectures. One possible direction to solve this is to have distributed learning agents. The associated challenges involve efficient handshaking of the agents and sharing the learning tables among multiple agents. Building a complete system to address the scalability problem is interesting, but challenging.

Scope   There are two aspects where the existing research works are lacking—heterogeneity of cores and thermal measurement. Most works on thermal management have focused on homogeneous cores. However, to extract performance and energy benefits, multiprocessor systems are equipped with heterogeneous cores. Examples include ARM's big.LITTLE architecture. The thermal behavior of different cores is different, even under the same workload. Additionally, not every thread or task can be mapped to every core. These challenges need to be factored in run-time thermal management policies. From the thermal measurements point, all the existing approaches have limitations—high simulation time for *HotSpot*, limited accuracy for *thermal gun* and recording speed for *thermal sensors*. One way to address the problem is to estimate the temperature from CPU performance statistics. Although some studies have been performed recently, accuracy is still an open problem.

Printed in the United States
By Bookmasters